The World Is Not 6000 Years Old—
So What?

The World Is Not 6000 Years Old—*So What?*

Antoine Bret

▲ CASCADE *Books* • Eugene, Oregon

THE WORLD IS NOT 6000 YEARS OLD—SO WHAT?

Copyright © 2014 Antoine Bret. All rights reserved. Except for brief quotations in critical publications or reviews, no part of this book may be reproduced in any manner without prior written permission from the publisher. Write: Permissions, Wipf and Stock Publishers, 199 W. 8th Ave., Suite 3, Eugene, OR 97401.

Cascade Books
An Imprint of Wipf and Stock Publishers
199 W. 8th Ave., Suite 3
Eugene, OR 97401

www.wipfandstock.com

ISBN 13: 978-1-62032-705-0

Cataloging-in-Publication data:

Bret, Antoine.

    The world is not 6000 years old—so what? / Antoine Bret ; foreword by Ian H. Hutchinson.

    xx + 108 p.; 23 cm—Includes bibliographical references and index.

    ISBN 13: 978-1-62032-705-0

    1. Religion and science. 2. Evolution (Physics). I. Hutchinson, I. H. (Ian H.), 1951–. II. Title.

QH366 .B74 2014

The Holy Bible, New International Version®, NIV® Copyright © 1973, 1978, 1984, 2011 by Biblica, Inc.® Used by permission. All rights reserved worldwide.

Manufactured in the USA.

Contents

Foreword by Ian H. Hutchinson vii
Note to the Reader xiii
Acknowledgments xv
Introduction xvii

1. If God Did Something, Does It Have to Be a Miracle? 1
2. Must Genesis 1 Be a Literal Account? 7
3. Some Misconceptions about Science 19
4. More than Six Thousand Years Traveling 45
5. Radio Dating and Astrophysics 61

   Conclusion 85

   Appendix A: A Little Atomic Physics 91
   Appendix B: A Little Nuclear Physics 97
   Bibliography 103

Foreword

Christians believe that God has spoken. He has spoken once and for all in the person of Jesus, the Word, whom we follow. God has also spoken, we Christians say, through the Bible. Shortly after the Reformation, the Bible became accessible within Christendom to practically everyone in their own language; and it has been by far the most widely read book ever since.

We take the Bible as the written word of God, and regard its teaching as the primary authority in matters of faith and life. It is worth wondering why. It is not hard to understand that Christians would treat the New Testament that way; because the Gospels are, after all, the source of our knowledge of what Jesus did and said and of the other events of his life, death, and resurrection. The New Testament also contains the insider history of the early years of the church, and the teaching of the Apostles: those who had been with Jesus in person.

It is less obvious, though, why Christians regard the Old Testament as authoritative. In part, it is because we believe God has been speaking through history in a special way to certain individuals, and to a particular nation, the Hebrews, whose history and spiritual development the Old Testament records. Most importantly, though, Christians take the Old Testament seriously for one main reason: because Jesus took it seriously. If that's so, then I suggest that our interpretation of the Old Testament ought to be modeled on his.

Jesus knew the scriptures well. The remarkable depth of his knowledge and understanding baffled the professional scribes in his youth. During his ministry, though, it became clear that his deep respect for the scripture was unorthodox by the standards of the religious teachers of the day. In fact, a great deal of Jesus's teaching in the Gospels consists of criticisms of the way that the scribes and the Pharisees interpreted and practiced the Old Testament. Even though Jesus affirmed the permanence of the Law, he and his disciples got a reputation for being lax in their observation of the Sabbath.

Though he cited scriptural passages by heart to contradict temptation, he would often eat with sinners, and undercut the religious leaders by sayings like "let him who is without sin cast the first stone."

He seemingly disregarded Biblical instructions about routine cleanliness but expanded them drastically in the direction of spiritual purity. The attitude of Jesus to the Old Testament was very far from being literalistic. It was ambiguous and often unsettling.

This ambiguity continued in the early church. Certainly their biggest internal debate was how much of the Old Testament law and ceremony new (non-Jewish) Christian converts should be required to observe. They had direct guidance on this matter in the form of visions and prophecies; but even so, establishing a consistent view of the role of Old Testament teaching in church life was a hard task. Successive Christian generations, down the centuries, have wrestled with the subtle and complex job of interpreting and applying the Old Testament to their current situation.

In addition to the Bible, Christians have from the beginning also considered God to have spoken in another way: through the natural world. Like the psalmists, Jesus himself frequently pointed to nature as providing signs of God's love and faithfulness. Metaphorically, it has been said that there are two books that reveal God: the book of God's word—the Bible—and the book of God's works—nature. This was almost a cliché for the founders of the scientific revolution in the seventeenth century; yet they meant it sincerely.

Seventeenth-century scientists, most of whom were Christians (and essentially all of whom were Europeans), recognized the possibility of contradiction between what the two books seemed to be saying. Apparent contradictions were present from the outset. For example, Copernicus's proposal that the earth orbits the sun rather than the sun orbiting the earth, brought forth arguments whose participants invoked the scriptures on both sides. The widely agreed principle, however, was that if science and the scriptures appeared to disagree, then one or other of them—either the science or the scriptures—needed to be reinterpreted.

Fast forward to twenty-first century America. Science has grown enormously in scope and influence, with technology spawned by it touching every facet of life. An uneasy tension between secular and religious worldviews has developed in society and even deep within many Christian denominations. The most dynamic branch of Christianity has embraced the power of technology, and is fascinated by science; yet within it has arisen a widespread doctrine to the effect that anything other than a literal "scientific" interpretation of the first few chapters of Genesis is a dangerous cop out to the secularists. This literal approach is generally referred to as Creationism. It grew up in the second half of the twentieth century. It was accompanied by various books that claimed to refute the accepted scientific descriptions of natural history, but whose arguments were rejected by almost all practicing scientists (whether religious or not).

There are different strands and shades of Creationism. One strand protests (understandably) against evolutionary metaphysics and its frequent use by anti-theists as an argument against God. This strand is willing to countenance an Earth that is very old, but not that animal species arose through common descent and natural selection. Accepting scientific evolution, they believe, opens the door to metaphysical evolutionism and is incompatible with God as creator. Besides, Genesis says God created each animal "after its kind." This book, though, is not about biology or evolution per se.

The most thorough-going Creationist strand believes the world must be just a few thousand years old, because its adherents think that the days of chapter one of Genesis are twenty-four hours each, and that the sequence of human history is described with literal scientific accuracy in the rest of the scriptures. It is this strand of belief that Antoine Bret addresses here: belief in a so-called Young Earth. He does so on two fronts. First, he discusses, from his own Christian believer's viewpoint, the problem of Bible interpretation. He explains how internal Biblical evidence indicates that some Bible passages must be interpreted non-literally. His aim is not to scorn the Bible or those who believe it; since he himself does. Instead he recounts straightforwardly the challenges, and some of the history, of the interpretation of Genesis. The early Church Fathers, who were ultimately responsible for the preservation and compilation of the Bible as we know it, recognized the need for Biblical interpretation that in some passages is not literal.

The second front Bret opens up is the science that leads us to say the universe is old.

It is quite difficult for non-scientists to assess the reasons why any scientific consensus is what it is, or how convincing is the evidence in its favor. There are certain scientific hypotheses, especially in cosmology, that are very uncertain and debatable. Yet, to the confusion of the public, they are all too often reported as if they are as well established as, say, the law of gravity. How does a non-specialist tell the difference?

Moreover, what science knows about the past often seems less certain than what it knows of the day-to-day matters of the laboratory and technology. Is the history of nature really as well known as the laws of nature? And if it isn't, how well do we know the universe is old?

As a practicing physicist, Bret well understands the science by which we can tell the age of the universe. He is also able to explain it to the non-specialist, and does so engagingly. In addition to the content of the science, he takes time to explain the processes of science. How scientific journals and the institutions of science operate, and how they serve to maintain the integrity of science. Conspiracies are extremely hard to maintain in the environment of the scientific literature. The system was designed to prevent us fooling ourselves or others.

Our scientific knowledge of the physical world is like a kind of tapestry woven of myriad different threads. Each thread stands for a series of experiments, observations, or theories. Some of the threads on their own are not very strong. They can be tested, doubted, and criticized; sometimes even broken. But together with all the other threads, they form a robust fabric in which each strand is supported by its neighbors. The self-consistent whole, the fabric, is far stronger than the individual strands. It is no exaggeration to say that all we know about the physical world combines to persuade us physicists that the universe and the earth are billions of years old.

We can't rule out logically or scientifically a final philosophical Young Earth refuge in which it is argued that the universe seems old only because God created it a few thousand years ago with the appearance of being much older. An almighty God has the power to do that. I would claim more than Bret about this philosophical recourse, though. It's true we can't rule it out logically; but I believe we can rule it out theologically; because, far from vindicating the truthfulness of God (the avowed aim of the literalists), the final philosophical refuge makes God into a deliberate deceiver. It defeats its own theological purpose.

The stakes are high. Will Evangelical Christians in America and elsewhere rediscover the compatibility of the Faith with real science properly understood?

Will individuals and institutions committed to the inspiration and trustworthiness of the Bible rediscover the power of scriptural interpretation that appropriately welcomes the presence of literary forms such as story, metaphor, and poetry, as well as literal history? Or will fear of unfaith insist on a narrow and ultimately unhistorical interpretation that has the appearance of doctrinal rigor, but little else to recommend it?

It is my hope that this book will make a contribution to relieving the fear and improving the understanding of the ways in which God has spoken—and still speaks.

Ian H. Hutchinson,
March 2014

Note to the Reader

The bibliography format for this book has been borrowed from the scientific literature and may require an explanation.

Books and scientific articles mentioned in the text are cited by means of a number between brackets—for example, [42]. The bibliography section at the end of the book then lists all the references cited, sorted in terms of their first author's last name. Entry [42] comes therefore forty-second in this order. It reads in the bibliography section:

[42] D. R. Gies and C. T. Bolton. The optical spectrum of HDE 226868 = Cygnus X-1. II spectrophotometry and mass estimates. *The Astrophysical Journal*, 304:371, 1986.

After the names of the authors and the title of the article, the name of the journal where the article was published comes in italic. Here, it is *The Astrophysical Journal*. The first number following the title, here "304," is the volume number of the journal where the article appeared. The second number, here "371," refers to the page number in that volume where the article can be found (even if you find something like "L21" instead of "21"). Finally, "1986" is, of course, the year of publication.

How to access these works? All the scientific journals mentioned in the bibliography have a Web page from which any article can be retrieved. While the abstracts of the articles are usually accessible for free, the full text is seldom free of charge.

An excellent shortcut to localize an article: just Google its title between quotes, and it should take you to the right place. For astrophysics articles, a free copy is usually available on the *arxiv.org* database.

The metric system of units is used throughout this book, as it is in the scientific literature.

Acknowledgments

Many thanks are due to the friends and colleagues who were willing to proofread the manuscript: Stephanie Bellamy, Guillaume Bellamy, Pablo de Felipe, Isabel de Sivatte, Dave Lawson, Jeff McClintock, Jean-François Mouhot, Stan Obenhouse, Kelly Petre, Lorenzo Sironi, and Rémy Vomscheid.

Introduction

I believe in God. I believe he's real and not a social or cultural construction. I believe in the Bible. I believe that both the Old and the New Testament are the inspired Word of God. I believe the Bible never contradicts itself and I have never found an apparent inconsistency that could not be resolved. I believe in Jesus Christ. I believe he existed physically, historically. I believe he was God in the flesh. I believe he performed miracles. I believe he died for my sins. I believe that his resurrection is an historical event and that, as Paul puts it in 1 Corinthians, Christianity is meaningless without it. I believe there will be a judgment day. I believe that both heaven and hell exist, and that neither will be empty. I believe all this, "for the Bible tells me so."

That is to say, this book is not intended to undermine the faith of anyone. I even hope it will strengthen the faith of many. I am a Christian, and I am also a scientist. When I became a Christian, one of the first questions I had to address had to do with the relation between the Bible and scientific knowledge. To my relief, I found I could perfectly harmonize both of them. I eventually assumed that if God created both the physical world and the Bible, there cannot be any contradiction between the two.

An NBC News poll conducted in 2005 focused on the question of the origin of human life.[1] To the question, "Which do you think is more likely to actually be the explanation for the origin of human life on Earth: evolution or the biblical account of creation?" 53 percent of respondents answered "the biblical account." Of these, 44 percent then emphasized they meant "God created the world in six days and rested on the seventh as described in the book of Genesis" rather than "God was a divine presence in the formation of the universe." This means many people have questions. *Many.* This means many were told in Sunday school that the world is six thousand years old and that confrontation with science will be, or already has been, a shock. This book is for them and for those willing to help.

---

1. See www.pollingreport.com/science.htm.

The structure of this book derives from its title. I do not think the world is six thousand years old, nor do I think this represents any threat to the Christian faith. From what I have observed, people tend to think science contradicts the Bible in at least two cases (there may be more, but I'll only treat two). The first case is when science finds a mechanism to explain an event that, according to the Bible, was provoked by God. A typical example is "God made the stars" versus the current theories of star formation from gravitational collapse of dust and gas in space. I will explain why, according to the Bible, an event can have been originated by God, while at the same time be perfectly understandable from the "mechanical" point of view. To be honest, this idea is far from original. Baruch Spinoza expounded it in 1670 in his *Theologico-Political Treatise*, and as a young Christian I was delighted to find he was biblically right.

I will then turn to the second case where science and the Bible seem to be in opposition: the interpretation of the book of Genesis. Let's state it plainly: I don't think Genesis 1 is a historical, literal account. The third-century church father Origen thought the same, and no one can suspect his faith was succumbing to the pressure of third-century science. There are many occasions when the Bible is obviously poetic, and that does not mean it is incorrect. It simply means it's wonderfully subtle.

I will thus start addressing the "so what" issue before dealing with the "six thousand years old" one. Why six thousand? Because if you consider the six days in Genesis to be real, literal, twenty-four hours days, you come up with a universe that young. I believe science has very good reasons to say that the universe is much older than that.

Before explaining a few of them (only a few), I will explain how science works, how science learns and how it comes to conclusions that can be considered definitive. I know "definitive" can seem far too strong here. I also know science progresses and that new theories replace older ones. I will nevertheless explain why "definitive" is definitely not out of context. I will also emphasize that scientists are *not* people who worry daily about proving the Bible wrong, nor are they committed to blocking whatever represents a threat to their theory.

The Berean-minded reader[2] may wish to go further than the content of this book. He may be keen on checking for himself how the observations reported in the science section were performed. I am not the one who measured the distance to such and such a star and found it greater than

---

2. Biblically, it's a compliment (Acts 17:11).

six thousand light years. The best I can do to bring the reader as close as possible to the scientific works I will mention is to tell him where to find the original articles written by those who did the measurements. I'm afraid this yields a quite obscure and unusual bibliography, but that is the price to pay for an improved trustworthiness. Still related to bibliographical issues, the reader will notice I'm not ashamed of citing Wikipedia. First, I have of course checked every single Wikipedia page I mention. Second, Wikipedia's articles are usually a very good introduction to the topic they deal with and include citations of a fair number of primary sources.

A warning may be needed: the science part of this book talks about math. There are even *equations*. Although there is absolutely no need to understand them, I do apologize to those who are completely unfamiliar with this language. The reason I included some key physical equations is simple: the concept of the laws of nature is central to this book. And these laws are mathematical. Science is all about trying to discern these laws. Writing a few of them under their mathematical form shows how *few* there are and how necessary must be their predictions. Unlike humans, particles follow their laws. As will be seen, these laws have been found to be valid *now*, and also *in the past*.

For improved clarity, the scientifically oriented chapters all start with a brief "In a Nutshell" section, summarizing the line of reasoning.

Simply put, the conclusion of this book is that there is no *natural* way to reconcile current astrophysical observations with a young, six-thousand-year-old universe. By "natural way" I mean a way accounting *only* for the known laws of nature. These laws have been found valid today and indeed explain an incredible amount of diverse phenomena. But there is more. It is frequently claimed that assuming these laws were valid in the past is no more than an assumption. But it is *not* an assumption. Thanks to astronomers, it is an *observation*. Our universe is so vast that observing far away is observing the *past*. Observations not only tell us about the present, they also tell us about the past. And what they tell us is that these laws, including the speed of light or the nuclear decay rates, were the same tens of thousands of years ago.

What is *not* ruled out, and indeed can hardly be ruled out, is the possibility of a miraculous intervention from God, in such a way that our universe only *looks* old, while it is not. Light from very distant galaxies may have been created in transit, decay rates miraculously accelerated, and so on. After all, water does not turn instantly into *good* wine, nor do fig trees wither in a minute. Science cannot tell anything about miracles, since by definition, they are events going *beyond* the known laws of physics. But if you think the physical world has behaved according to these laws, then a young universe is ruled out. This is all I will argue.

I perfectly understand that *many* questions can arise when believers undertake to learn the teaching of modern science. After all, most people don't really know how their microwave oven works. This means that, in general, the opinion that the universe is old eventually relies on . . . *trust*. Trust in some TV programs, trust in the newspapers, trust in a few good pop-science books, trust in some remote high school or college memories. At any rate, few can claim they really know *why* they think what they think. Isn't it then normal to start doubting this all, when trust in the Bible steps in? How does a person avoid thinking, "I don't know what a galaxy is, or a light year, and all this stuff. But Genesis 1, I can read myself"? Indeed, I admire the courage to admit one's doubts, in front of the "company of mockers" who hardly know better. My hope is that this book will answer questions and help the seeker find peace.

Centuries ago, the author of Ecclesiastes wrote, "Is there anything of which one can say, 'Look! This is something new'? It was here already, long ago, it was here before our time" (1:10). Though I hope it is the case, I'm not sure you will find here something never before read. I simply wish to share the reasons I don't see science as a threat to my faith, and why I do believe the world is older than six thousand years. I hope it will be useful to you.

1

If God Did Something, Does It Have to Be a Miracle?

A man sits on a rooftop in a flood, refusing three rides from passing boats as he waits for the Lord to provide. Water keeps rising, and he drowns. He gets to heaven and asks God why he was forsaken. "But I sent three boats!" God replies.

"God causes his sun to rise" (Matt 5:45).
Who said it is because the earth rotates?

God created the heavens and the earth. God said, "Let there be light." God said, "Let the land produce vegetation." God made two great lights; He also made the stars.

It can be difficult to reconcile the opening verses of Genesis with any sort of scientific explanation or mechanism. At the heart of the conflict with science lies the vivid sense that if God is responsible for something, it shouldn't be possible to find a mechanism explaining how this "something" came to be. If the Bible says "God did something," it must *remain* unexplained.

Surprisingly, most believers and atheists seem to *agree* on this point: if some natural process can explain an event, then God does not have anything to do with it. In an interview with *Newsweek*, Nobel physicist Steven Weinberg claimed, "As science explains more and more, there is less and less need for religious explanations."[1] Sadly, many believers would agree with him.

A historical anecdote shows this mindset is not new. About two centuries ago, the French astronomer and mathematician Pierre-Simon Laplace (1749–1827) had a conversation with Emperor Napoleon I about his book on celestial mechanics. Victor Hugo records their conversation (reference [50], page 453):

> Someone had told Napoleon that the book contained no mention of the name of God. Napoleon, who was fond of putting embarrassing questions, received it with the remark, "M. Laplace, they tell me you have written this large book on the system of the universe, and have never even mentioned its Creator." Laplace, who, though the most supple of politicians, was as stiff as a martyr on every point of his philosophy, drew himself up and answered bluntly, "Sir, I had no need of that hypothesis."

Today, many would find their faith threatened should biologists discover a way DNA can be naturally synthesized, or physicists formulate a theory explaining how the Big Bang occurred. What I would like to show in this section is that the premise that any natural explanation means that God was not involved is *biblically* flawed. According to the Bible, God does not need any gap of understanding to have a chance to act. A famous quote by Albert Einstein perfectly illustrates my point; of the scientist, he says,

> His religious feeling takes the form of a rapturous amazement at the harmony of natural law, which reveals an intelligence of such superiority that, compared with it, all the systematic thinking and acting of human beings is an utterly insignificant reflection.

To me, the core of this claim is that God may be found in the very natural laws we understand, and not only in the supernatural events we can't even remotely grasp. God's role does not stop when the textbook starts. He is in the textbook as well. You don't need any scientific gap to get a chance to find God. He's already on both sides of the gap! Surprisingly, this is just *biblical*.

---

1. "In Search of the God Particle," *Newsweek*, March 23, 2008, http://www.thedailybeast.com/newsweek/2008/03/23/in-search-of-the-god-particle.html.

God Sent Me Ahead of You

The book of Genesis goes to great lengths to explain how Joseph came to Egypt and became a very prominent person there. We are told he was the first son of Rachel, and thus he was favored over all his brothers. Out of jealousy, they wanted to kill him, but instead they sold him to merchants going to Egypt. His story in Egypt is also told in detail, so that we understand perfectly all that happened to him prior to his audience with his brothers in Egypt. Yet, in Genesis 45:5, he claims, "It was to save lives that God sent me ahead of you." Later, in verse 8, he adds, "So then, it was not you who sent me here, but God." Clearly, for him, *God* brought him to Egypt. He had certainly not forgotten the years of trial and his journey up to that point, but for him it is obvious: God made it all happen.

Now, what if the Bible had no record of how Joseph came to Egypt, while retaining *only* Genesis 45:5–8? All we would be told is that God brought Joseph to Egypt. And what if some archaeologists discovered an inscription telling the story of a young stranger who ruled Egypt for some years after being taken there as a slave? Many would argue that this young stranger may well have been Joseph, while others would claim it is impossible because the Bible says that God, and no one else, brought Joseph to Egypt.

Some would find this archaeological discovery a threat to their faith only because they would not be able to reconcile "God did this" with the possibility of an explanation. Others would see there a proof of the fallacy of the Bible since it claims that God was responsible for Joseph's coming to Egypt, whereas the archaeological record shows he did it another way. But both would be wrong. Indeed, the Bible does tell us God did something while giving all the details of how it came to be.

He Turned Their Hearts to Hate His People

Let us now turn to Exodus 1:6–11:

> Now Joseph and all his brothers and all that generation died, but the Israelites were exceedingly fruitful; they multiplied greatly, increased in numbers and became so numerous that the land was filled with them. Then a new king, to whom Joseph meant nothing, came to power in Egypt. "Look," he said to his people, "the Israelites have become far too numerous for us. Come, we must

deal shrewdly with them or they will become even more numerous and, if war breaks out, will join our enemies, fight against us and leave the country." So they put slave masters over them to oppress them with forced labor, and they built Pithom and Rameses as store cities for Pharaoh.

We are told how the Israelites were progressively brought to slavery, and we perfectly understand how an early anti-Semitism developed as they became increasingly numerous and powerful, while the Egyptians forgot that Joseph had saved them. On the other hand, Psalm 105:24–25 simply says, "The Lord made his people very fruitful; he made them too numerous for their foes, whose hearts he turned to hate his people, to conspire against his servants."

Here again we discover the same pattern; on the one hand, we find a sociological explanation of the situation, while on the other hand, the Bible claims that God made it happen. We can once again imagine what would happen if the explanatory part of Exodus were missing in the Bible and discovered elsewhere. And once again, we find this would be considered by some believers a threat to their faith, while others would see there a proof of the fallacy of the Bible.

I Have Set My Rainbow in the Clouds

Genesis 9:13—"I have set my rainbow in the clouds"—makes it very clear that God is directly responsible for the rainbow. Does that mean it is a supernatural event? Circa 78 AD, the Roman philosopher Pliny the Elder wrote,

> What we name Rainbows frequently occur, and are not considered either wonderful or ominous; for they do not predict, with certainty, either rain or fair weather. It is obvious, that the rays of the Sun, being projected upon a hollow cloud, the light is thrown back to the Sun and is refracted, and that the variety of colors is produced by a mixture of clouds, air, and fire. The rainbow is certainly never produced except in the part opposite to the Sun, nor even in any other form except that of a semicircle.[2]

You see, the mechanism behind the rainbow has been known for centuries. Does that mean God's hand is ruled out? I don't think so. I just think

---

2. *Natural History*, Bk. 2, Chap. 60.

God allows us, from time to time, to understand his plans in part. If the recipe for a rainbow is "sun plus rain," Jesus confirmed in the Sermon on the Mount that God directly produces these events (see below).

From My Mother's Womb

Psalm 71:6 reads, "From birth I have relied on you; you brought me forth from my mother's womb. I will ever praise you." The psalmist is clear: God brought him forth from his mother's womb. Did he have a miraculous birth, or should we deduce God brought us all from our mother's womb? I think the psalmist well knows that he was born like every one of us, and yet he is also perfectly aware that, indeed, God has always been the one behind his birth. Who, based on this verse, would claim that midwives, nurses, doctors, and hospitals don't exist?

As the Rain and the Snow

Isaiah 55:10–11 declares,

> As the rain and the snow come down from heaven, and do not return to it without watering the earth and making it bud and flourish, so that it yields seed for the sower and bread for the eater, so is my word that goes out from my mouth: It will not return to me empty, but will accomplish what I desire and achieve the purpose for which I sent it.

These verses are particularly interesting because they draw a clear parallel between the way God's word works and the way some natural processes work. The cycle of water described here has been known for centuries. It is very interesting that here God compares the action of his word to something supremely "natural."

The Sermon on the Mount

On repeated occasions in the Sermon on the Mount, Jesus explicitly claims God's authorship of routine events: "I tell you, love your enemies and pray for those who persecute you, that you may be children of your Father in heaven. *He causes his sun to rise* on the evil and the good, *and sends rain* on the righteous and the unrighteous" (Matt 5:44–45, emphasis mine). Doesn't the sun rise because the earth rotates? And doesn't the rain fall because water condenses into clouds under certain conditions? So, was Jesus wrong? Clearly not.

Look now at another verse: "Look at the birds of the air; they do not sow or reap or store away in barns, and yet *your heavenly Father feeds them*. Are you not much more valuable than they?" (Matt 6:26, emphasis mine). The feeding of a bird is no mystery at all (ask the earthworm). Is it possible to find a process more natural than this one? Yet, Jesus is clear: God is the one who feeds them. Later in the Sermon, Jesus declares, "Are not two sparrows sold for a penny? Yet not one of them will fall to the ground *outside your Father's care*" (Matt 10:29, emphasis mine).

Who would claim that the fall of a sparrow is a miracle? Who would pretend that science's teaching on the sun's rising or the rain's falling is wrong, because Jesus said God *causes* both of them? And who would argue that sparrows are fed supernaturally? Jesus knew that these events were familiar to his listeners (this is why he chose them to illustrate his message). Yet he explicitly points to God as the author of each one of them. Jesus wouldn't agree that something no longer needs God once it is explained.

If Christianity had to die the day something God is supposed to do was explained "without God," it would have died long ago! Any sunrise would be enough. So why fear scientific explanations? God does not have to be—and he is not—a "God of the gaps," hoping for some event that science definitely *cannot* explain in order to be needed. As stated earlier, the biblical message is that he's already on both sides of any gap.

2

Must Genesis 1 Be a Literal Account?

Is Genesis 1 to be taken literally? The question lies at the heart of this book. If Genesis 1 is to be considered at face value—if the seven days are seven twenty-four-hours days—then you can compute the age of the universe based on the Bible. During the seventeenth century, Irish Archbishop James Ussher gave 4000 BC as an approximate date. But reading Genesis literally is not the only option for someone who claims to believe in the Bible.

Nonliteral Reading in Early Christian Literature

To start with, we have writings from the early church that would *not* consider the Genesis account to be literal. This does not mean that Genesis is not literal, nor does it mean they were right. But it does mean that some sincere Bible believers read Genesis 1 symbolically at a time when there was no science pointing to an old universe. In other words, they weren't doing so because they felt compelled to in order to make it consistent with their modern science.

Justin Martyr

Justin Martyr was born in Palestine in about 100 AD. He was executed for his faith in Rome in 165 AD. When commanded to worship Rome's pagan gods, he refused. He was then told, "If you do not obey, you will be tortured without mercy," to which he replied, "That is our desire, to be tortured for Our Lord, Jesus Christ, and so to be saved, for that will give us salvation and firm confidence at the more terrible universal tribunal of Our Lord and Savior." Here is what Justin wrote about the fact that although Adam was told he would die the day he ate of the forbidden fruit, Genesis says he lived almost one thousand years:

> For as Adam was told that in the day he ate of the tree he would die, we know that he did not complete a thousand years. We have perceived, moreover, that the expression "The day of the Lord is a thousand years" is connected with this subject.[1]

Another writing of his is worth quoting from here ([9], page 611):

> And the fact that it was not said of the seventh day equally with the other days, "And there was evening, and there was morning," is a distinct indication of the consummation which is to take place in it before it is finished.

We thus find Justin had a symbolic reading of Genesis. His point in the first quote is that Adam didn't die when he ate the fruit of the tree, although God warned him "in the day that thou eatest thereof thou shalt surely die" (Gen 2:17, KJV). Which sort of "day" is that? Not only did Adam not die as soon as he ate the fruit, but Genesis 5:5 claims that he lived 930 years. For Justin, there was only one way to reconcile "you will die the *day* you eat the fruit" with "Adam lived 930 years": the *day* being considered here was a thousand-year day, not a twenty-four-hour one.

Justin therefore concluded that Adam both ate the fruit *and* died during the seventh "day," even if some 930 years passed between these two events. He reiterates this specificity of the seventh day in the second quote, noting the refrain "there was evening, and there was morning" is missing only for this very day. The same idea of a long seventh day is found in the writings of Victorinus (see below), but let us turn to Origen first.

Origen

Origen was born in Alexandria in 185. At the age of seven, he saw his father beheaded for being a Christian. According to the early church historian Eusebius, Origen literally castrated himself to avoid carnal temptation. When the persecution of the Roman emperor Decius broke out in 250 AD, he was sent to jail and tortured. I wouldn't call him a lukewarm Christian. During this time, he wrote *Contra Celsum* to defend Christianity. He never

---

1. *Dialogue with Trypho the Jew.*

recovered from the torture he suffered and died three years later, circa 254 AD. Yet he wrote these words on Genesis 1:

> For who that has understanding will suppose that the first, and second, and third day, and the evening and the morning, existed without a Sun, and Moon, and stars? And that the first day was, as it were, also without a sky? And who is so foolish as to suppose that God, after the manner of a husbandman, planted a paradise in Eden, towards the east, and placed in it a tree of life, visible and palpable, so that one tasting of the fruit by the bodily teeth obtained life? And again, that one was a partaker of good and evil by masticating what was taken from the tree? And if God is said to walk in the paradise in the evening, and Adam to hide himself under a tree, I do not suppose that anyone doubts that these things figuratively indicate certain mysteries, the history having taken place in appearance, and not literally.[2]

Note the style here, reminiscent of James 2:20: "You foolish person . . ." Origen does not simply suggest Genesis might be symbolic. He straightforwardly declares that no one who "has understanding" should take these verses "literally." What is important here is not whether he was right or wrong to write this way. What is important is that nearly two thousand years ago, some faithful Christians already had a symbolic reading of these chapters.

Cyprian of Carthage

Cyprian of Carthage died in 258 AD. His birthdate is not known. An eminent and wealthy man of Carthage, he became a Christian late in life, being baptized in 246 AD. He became bishop of Carthage in 249. In 256 began the persecution of Christians by the Roman emperor Valerian. Cyprian was sent to jail in 258. Upon learning that he had been sentenced to death, he replied only, "Thanks be to God!" He was beheaded the same year, after he refused to worship Roman idols. Among his writings can be found the following words, strongly suggesting that he counted each day in Genesis as one thousand years: "As the first seven days in the divine arrangement containing seven thousand of years . . ."[3] Cyprian is quite clear here: for

---

2. *De Principiis* 4:16.
3. *Exhortation to Martyrdom*, 11.

him, the first seven days do *not* comprise seven twenty-four-hour periods, but seven thousand years.

Victorinus

Victorinus died circa 303 AD. He was bishop of Pettau, now in Slovenia. He died as a martyr under the persecution of the Roman emperor Diocletian. Among his few writings that have been recovered is a commentary on Genesis in which one finds the following words:

> Who, then, that is taught in the law of God, who that is filled with the Holy Spirit, does not see in his heart, that *on the same day* on which the dragon seduced Eve, the angel Gabriel brought the glad tidings to the Virgin Mary.[4]

We find here that Victorinus thought that the dragon seduced Eve on the very *same* day Gabriel visited Mary. This is very reminiscent of the quote by Justin Martyr given above. Of course, it implies that these authors did not consider the seventh day to be a literal, twenty-four-hour day.

Note again the tone: "Who ... that is taught in the law of God ... filled with the Holy Spirit, does not see in his heart ..." The words of Origen and Victorinus indicate that not only was there a debate even then, but also that some had taken a very strong stance on an interesting position. Today, many believe a literal reading of Genesis is the only spiritual option. Nearly two thousand years ago, Origen and Victorinus argued instead that only the nonliteral interpretation is spiritual.

Can we find early *literal* readings of Genesis 1? Of course, yes. But it is very interesting to discover that there were conflicting views. There may be other nonliteral statements about Genesis in other early Christian writings, such as *The Epistle of Barnabas*, for example. But their number is not important. What is important is that some sincere Christians had a nonliteral reading of Genesis long before science had anything to say about the age of the cosmos. Should we then feel that anything short of a literal reading of Genesis 1 constitutes unbelief? It seems the answer is no.

---

4. *On the Creation of the World.* Italics mine.

When the Bible Itself Is Symbolic

Let me now turn to some verses that demonstrate that the Bible is sometimes symbolic. Does this mean the Bible cannot be trusted? Absolutely not. It means that besides being the word of God, a book of incomparable wisdom and an account rooted in history,[5] the Bible is also a great piece of literature and poetry.

Of course, there are places in the Bible where poetry and symbols are perfectly expected, as in the Psalms. Psalm 18:6–8, for example, reads, "In my distress I called to the Lord. . . . Smoke rose from his nostrils, consuming fire came from his mouth . . ." I hope no one would deduce from these verses that God has a mouth and a nose with nostrils (and that both are fireproof). This is clearly poetry. Besides explaining why I think Genesis 1 should be read symbolically, I will mention other verses in which the Bible is not to be taken literally, although the context does *not* make it obvious.

In Genesis

- This first item was already outlined by Origen nearly eighteen hundred years ago: since the sun was created on day 4, how could there have been a twenty-four-hour day before? Of course, it could have been so, but the point here is that the text itself doesn't make it mandatory.

- Genesis 1 does not start with the first day, but *before*. Verses 1–3 read, "In the beginning God created the heavens and the earth. Now the earth was formless and empty, darkness was over the surface of the deep, and the Spirit of God was hovering over the waters. And God said, 'Let there be light,' and there was light."

    We thus have the heavens and earth being created at the "beginning." *Then*, and only then, God creates light and calls it "day" in verse 5. Where is it written that the Spirit hasn't been hovering over the waters for twenty billion years, before God created light? Since verse 5 explicitly relates light to a day, how can there be any day *before* light is created?[6]

---

5. Visiting the Louvre in Paris, or the British Museum in London, with a Bible in hand is a great experience!

6. I know the so-called gap theory relates to this idea. Regardless of the sticker you put on it, the bottom line remains: the first day doesn't start with Genesis 1:1.

- Genesis 2:20 says that Adam had *one* day to name "all the livestock, the birds in the sky and all the wild animals." There are 10,000 species of bird. Regarding the livestock and the wild animals, let us focus on mammals only. There are 5,500 species of non-marine mammals alone (Adam didn't have to name "the great creatures of the sea and every living thing with which the water teems and that moves about in it"). Adam, then, had twenty-four hours in which to name at least 15,500 animal species. That means naming a species every 5.57 seconds, providing he didn't do anything else during those twenty-four hours.[7]

  The problem is that on the *very same day*, "God caused the man to fall into a deep sleep; and while he was sleeping, he took one of the man's ribs and then closed up the place with flesh" (Gen 2:21). It had to be the very same day because Genesis 1:27 says that on day 6 "male and female he created them." In addition, God began day 6 creating "the wild animals . . . the livestock according to their kinds . . . and all the creatures that move along the ground."

  Granted, we can imagine that God first created the animals in an instant, and then that Adam's "deep sleep" lasted just a few seconds, after which he spent a little less than twenty-four hours (nonstop) naming a species every five seconds (and here we are counting just the birds and mammals). But doesn't such a scenario require much more imagination than a symbolic reading?

- Which kind of trees are the "tree of life and the tree of the knowledge of good and evil"? Surely, they bear *symbolic* fruits, but which kind of *literal* fruits should they bear?

- Genesis 3:7 says that after Adam and Eve ate the forbidden fruit, "the eyes of both of them were opened." Were they literally blind before? It would have been difficult for Adam to name so many species of animal without seeing them. Also, which kind of God would make "trees that were pleasing to the eye" (Gen 2:9) for a blind man? What is more, Genesis 3:6 says that Eve succumbed to the temptation when she "*saw* that the fruit of the tree was good for food and *pleasing to the eye*" (emphasis mine).

---

7. Note that the number of species in question cannot be easily deduced because the Bible's definition of a species, "reproduce after their own kind," fits quite well the one of modern biology, "a group of organisms capable of interbreeding and producing fertile offspring." See Wikipedia, "Species," or reference [93], page 93.

It is clear that Adam and Eve could see, physiologically speaking, before Genesis 3:7. This verse must therefore be symbolic, meaning "the [spiritual] eyes of both of them were opened" when tasting the forbidden *spiritual* fruit.

- Genesis 3:24 says that after God drove the man out of Eden, "He placed on the east side of the Garden of Eden cherubim and a flaming sword flashing back and forth to guard the way to the tree of life." According to Genesis 2:14, the Garden bordered the rivers Tigris and Euphrates, which can still be seen today in the Middle East. If the cherubim and the sword are literal, where are they? Millions of people have gone through this region during the last centuries. Satellites are now mapping the globe with a resolution of less than one meter. The cherubim and flaming sword would have been found by now if they had existed physically and literally.

  Maybe the cherubim were wiped out by the Flood. But why are the rivers Tigris and Euphrates still there, and not the cherubim?

- According to Genesis 19:24, "The Lord rained down burning sulfur on Sodom and Gomorrah". In Jude 1:7 we read that "Sodom and Gomorrah . . . are set forth for an example, suffering the vengeance of *eternal* fire" (KJV, emphasis mine). We know where Sodom and Gomorrah were approximately, and it is clear that the place no longer burns today.

In the Rest of the Old Testament

- Isaiah 11:12 claims that God "shall assemble the outcasts of Israel, and gather together the dispersed of Judah from the *four corners of the earth*" (KJV, emphasis mine). It is easy for us to claim that this verse is clearly nonliteral, since we all know that the earth does not have four corners. But it is equally obvious that if such verses are still interpreted literally by some *today*, many sincere and faithful readers of past centuries must have thought that the earth was flat because of it.

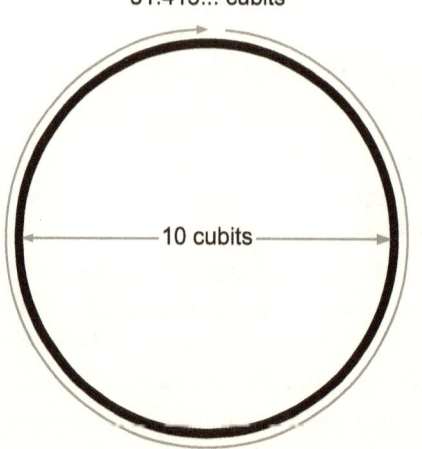

Figure 1: According to 1 Kings 7:23, the circumference of a circle of diameter 10 cubits measures 30 cubits. In reality, it is 10π = 31.4159... cubits. This verse cannot be literal, although nothing in the context tells us so.

- At about the time King Solomon built his palace, 1 Kings 7:23 tells us: "He made the Sea of cast metal, circular in shape, measuring ten cubits from rim to rim and five cubits high. It took a line of thirty cubits to measure around it." If taken literally, that means the circumference of a circle of diameter ten cubits is thirty cubits. We thus find the ratio circumference/diameter = 3. But we all know this ratio is π (3.14159 ...), so that the "Sea of cast metal" must have been some 31.41 cubits, as shown in figure 1. That doesn't mean the Bible is wrong. It just means this verse is not to be taken literally, even if nothing in the context suggests it.

- First Chronicles 16:30 says this about God: "Fear before him, all the earth: the world also shall be stable, that it be not moved" (KJV). If this verse is literally true, how can the earth turn around the sun?

  Granted, Job 26:7 reads, "He spreads out the northern skies over empty space; he suspends the earth over nothing," so that the earth *can* move. I've often heard Bible believers invoke this verse, claiming the Bible knew long ago about gravity and so on. So 1 Chronicles 16:30 must be poetry. But suppose you're a second-century Christian who *never* saw a picture of the earth taken from the moon: how would you know which verse is poetry and which is not?

  Last but not least, Job is generally classified among the *poetic* books, and Chronicles among the *historical* ones. Yet, the symbolic verse about the immovable earth is in Chronicles, and the literal one stating that the earth is suspended over nothing appears in Job. If you were living three hundred years ago, how would you tell, from the Bible alone, that the earth turns around the sun?

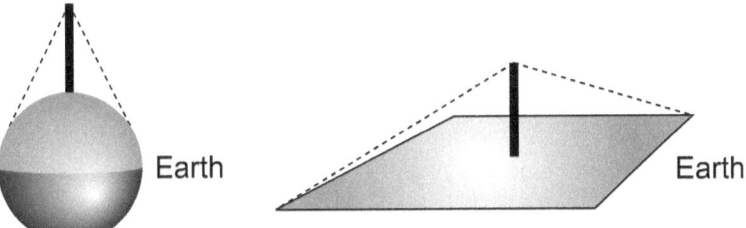

Figure 2: No matter how high an observer is on the upper hemisphere, he cannot be seen from the lower one (*left*). A literal reading of Daniel 4:10–11 or Luke 4:5 demands a flat earth (*right*).

- Daniel's vision in Daniel 4:10–11 is possible only if the earth is flat: "I saw a tree of great height at the center of the earth; the tree grew and became strong, reaching with its top to the sky and visible to the earth's farthest bounds." On a spherical earth, a tree, no matter how tall it may be, cannot be seen from the other hemisphere (see figure 2). We will explore this further in the next section.

In the New Testament

- Luke 4:5 says that when Jesus was tempted in the desert, "The devil led him up to a high place and showed him in an instant all the kingdoms of the world." As in Daniel 4 above, that would only be possible if the earth were flat. On a spherical earth, an observer can see no more than his own hemisphere (see figure 2).

    Here again, one could "counterattack" by claiming that Isaiah 40:22—"He sits enthroned above the *circle* of the earth"—implies a spherical earth. So, those verses in Daniel and Luke must be poetry. Again, how could you tell, without knowing it in the first place? Here again, it is not easy to tell from the context, as Luke is definitely a *historical* book.

    My point is not to prove that the Bible contradicts itself. My point is that the Bible is often symbolic, allegoric, or poetic, even when not obvious from the context. How unfortunate it is to lose the beauty of it over endless, pseudoscientific discussions about which verse means what.

- Acts 2:5: "Now there were staying in Jerusalem God-fearing Jews from *every nation under heaven*." Some of them came from China?

    Colossians 1:23: "This is the gospel that you heard and that has been proclaimed to *every creature under Heaven*." Did Paul preach in Australia as well?

    Finally, although this is the New Testament section, let me quote Genesis 41:57: "*All the world* came to Egypt to buy grain from Joseph, because the famine was severe everywhere." People came even from America?

    1 Kings 10:24: "*The whole world* sought audience with Solomon to hear the wisdom God had put in his heart". Every single Chinese wanted to be there?

    It is frequently argued that the days of Genesis *must be* twenty-four-hour periods, because the same word is repeatedly used with this very meaning in the Old Testament. But the same could be said about the Hebrew word for "world"[8] in Genesis 41:57 or 1 Kings 10:24. It is the same Hebrew word you find for "earth" in Genesis 1:1: "In the beginning God created the heavens and the *earth*." Therefore, if "earth" in Genesis 1:1 refers to the whole globe, then "world" should refer to the whole globe as well in Genesis 41:57 or 1 Kings 10:24. Should we necessarily believe a few Australians came to Solomon? I don't think so.

    Let us return to the New Testament. Could "every nation under heaven" in Acts 2:5 mean something else to a first-century reader? Would any scholar claim that the Greek words for "under heaven" strictly meant, in Roman times, *circa mare nostrum*, as Romans would have called the Mediterranean region? To our first-century reader, these words implied that the gospel had been preached to every single living human being. Yet, Paul never traveled to Australia, and carbon-14 dating (see chapter 5), which correctly dated the Dead Sea Scrolls[9] to a few centuries BC, reveals that some people lived there at that time, as well as in North and South America and Asia.

    Should we thus claim that the Bible is wrong because it says the gospel was preached to "every nation under heaven" in the first

---

8. This is *'erets*, Strong number 776.

9. A collection of Old Testament fragments found around 1950 near the Dead Sea, in Israel. Dated by the 14C technique to a few centuries BC, they pushed back the date of the oldest known Old Testament original manuscript by more than one millennium! Indeed, 14C and the Bible are best friends. See Wikipedia, "Dead Sea Scrolls."

century, although this is not literally true? Should we conclude that the Bible is a fairy tale because there were humans on earth who did *not* seek an audience with Solomon? Of course not. So why should the days of Genesis be understood as twenty-four-hour time periods, even if the Hebrew word for "day" usually means this?

Symbolic Verses . . . or Not?

I will end this part with two instances in which the Bible could very well be literal, or could very well be symbolic.

- In Exodus, we are told that the Israelites left Egypt. Moses leads them from the Red Sea, and in Exodus 15:23 they come to Marah, where they cannot drink the water. Then, in verse 27, they come to Elim, where "there were *twelve* springs and *seventy* palm trees, and they camped there near the water" (emphasis mine).

    How biblical are these numbers! Joseph came to Egypt with his twelve brothers and a total of seventy people (Gen 46:27). Jesus had twelve disciples before he appointed seventy-two (NIV) or seventy (KJV) others (Luke 10:1). The pattern is the same: there is one "precursor," then twelve and seventy first- and second-generation followers, respectively. In Exodus 15, the water (precursor) feeds the twelve springs (first generation), which in turn feed the seventy palm trees (second generation). Remember, Paul explicitly relates the water that the Israelites drank in the desert to Christ (1 Cor 10:4). Maybe there really were twelve springs and seventy palm trees, and God made the physical number coincide with the symbols. But could we say that the Bible is erroneous if there were only ten springs and fifty palm trees? I don't think so. Jesus, the Word made flesh, taught at length in parables. Can't the Word itself do the same?

- In the Gospel accounts of Jesus' baptism, we read the following:

    > As soon as Jesus was baptized, he went up out of the water. At that moment heaven was opened, and he saw the Spirit of God descending *like a dove* and alighting on him. (Matt 3:16)

    > Just as Jesus was coming up out of the water, he saw heaven being torn open and the Spirit descending on him *like a dove*. (Mark 1:10)

> When all the people were being baptized, Jesus was baptized too. And as he was praying, heaven was opened and the Holy Spirit descended on him *in bodily form like a dove*. (Luke 3:21–22)
>
> Then John gave this testimony: "I saw the Spirit come down from heaven *as a dove* and remain on him." (John 1:32)

Did an actual dove literally descend upon Jesus, or did the Holy Spirit indeed descend on Jesus, only *like* a dove, so that we are left to our own imagination regarding what actually happened? I think it would be regrettable to refuse the symbolic and miss a wonderful bigger picture. Throughout the Bible, God makes the plans, Jesus establishes them, and the Holy Spirit maintains what has been established. It's like building a skyscraper: first, the architect plans it, then the builder erects it, and finally, the custodian maintains it. God made the plans of the ark (Gen 6:14–16); Noah, a type of Jesus,[10] built it; and a dove, the Holy Spirit, led it (Gen 8:9–12). God made the plans for the Church (Matt 25:34; Eph 1:4), Jesus built it (Matt 16:18), and the Holy Spirit led it (John 14:16, 26). And the pattern repeats endlessly.

So, was the dove of Jesus' baptism physical or symbolic? It could be either. Would the Bible be proved wrong if no dove had descended to alight on Jesus? Absolutely not.

Some of the verses I just mentioned must be nonliteral (π is not 3), even when the context doesn't make it obvious. Others are likely to be so, and still others *may* be so. We also see that there are cases in which two verses seem to contradict each other, and it is not easy to tell from the Bible alone which one is poetry and which one is not. At any rate, the reliability of the Bible is never at stake—only the reliability of its interpreter.

---

10. "The word 'type' is generally used to denote a resemblance between something present and something future, which is called the 'antitype'" (M. G. Easton, *Easton's Bible Dictionary*).

3

Some Misconceptions about Science

In a Nutshell

This chapter is designed to clarify several questions or misconceptions, frequently arising in this kind of debate. First, a few important issues regarding the laws of nature:

- A *very* frequently asked question: "You say the laws of physics forbid such and such conclusions. But aren't these laws condemned to become outdated some day? Isn't science always progressing?" Yes science progresses by elaborating new laws. But new laws have to *contain* the old ones. They should obviously explain something *more* than the old laws. But they must also explain *everything* the old laws already explained. The example of gravity illustrates very well this process.
- A given theory is therefore valid within the limits of its *validity domain*. Each new theory contains the former, with an increased validity domain.
- The physics involved in dating techniques does so *within* its validity domain. Such conclusions do not rely on shaky extensions of what is known. They rely on well-established results that any further theory will necessarily have to reproduce.
- How can so few equations be the guide to so many phenomena? A key is in the strict "supervisors" who oversee the development of any new theory: coherence with logic and with observation. They make it difficult, but they make it reliable as well.

Next, a few misconceptions regarding the attitude of the scientific community:

- The scientific community does *not* live to prove the Bible wrong. With the exception of Richard Dawkins, who indeed appears to do so, millions of scientists (including me) wouldn't even list this goal among their top twenty work priorities.
- Physicists do *not* fear their theories may be proved wrong. It's rather the contrary: known laws are constantly put to the test in order to get a chance to find the next ones.
- The scientific community is *not* a select private club, where unexpected results have no chance at all to be heard. I will explain how the scientific literature system works. History shows it is ready to welcome big surprises, as long as they are consistent with logic and observation.

I don't think science necessarily has to do with wearing glasses or being in a lab; or knowing words that only a few others understand; or being an atheist, an agnostic, or a whatever-ist; or using math and a big computer; or having a PhD; or writing books and articles and giving talks in conferences; or anything usually associated with science and scientists. The best, and by far the simplest, definition of science I know comes from Richard Feynman, one of the most fascinating physicists of the twentieth century (reference [36], page 205): "Science is a way of trying not to fool yourself." Short, isn't it? But to me, it captures the essence of science: trying to guess how things work, and then finding a clear-cut way to check that guess. Or asking a question, and trying to find a way to answer it *from the facts*, instead of talking about it over and over.

Let me choose, *purposely*, an example completely unrelated to physics: How many people would *willingly* sign up to become a slave at the time of the transatlantic trade? Who on earth would make such a move? I can perfectly imagine endless discussions around it. What would a *historian* interested in the answer do? He could try to find written registers of slaves and look for any mention of such behavior. That's exactly what the American historian David Galenson did [41] (cited in reference [72], page 73). Patiently going through some twenty thousand contracts archived in London, he came to the conclusion that about 6 percent of those deported between 1654 and 1775 were *volunteers*.

One can have a hard time believing it, wonder about it, even disagree—but on what ground? That may not be the exact number, but this is the best guess we can make. Is there a better way not to fool myself than going to the archive and counting the number of records indicating "volunteer"?

This example may highlight what science is. After all, it is just a way of trying not to be fooled. In the same way that such a simple definition of a big word like *science* may have surprised you, the topics covered by the following sections may be unexpected. But I really think they're instrumental in getting a better understanding of what science has to tell us about the age of the world.

How Do New Theories Replace Old Ones?

False ideas abound, too, about what is "right" or "wrong" in physics. What do scientists mean when they claim a theory is "wrong"? How do new, "right" theories replace old, "wrong" ones? The history of the theory of gravitation offers a great opportunity to understand these issues.

At the beginning of the seventeenth century, Kepler observed that planets were orbiting around the sun following geometrical figures called "ellipses," which are sort of flattened circles. By the end of the century, Newton discovered that everything Kepler had observed (there was more than the elliptical motions) could be derived from the following mathematical laws:

$$F = G \frac{M_1 M_2}{r_{1,2}^2} \qquad (1)$$

$$v(t_2) = v(t_1) + \frac{F}{M} (t_2 - t_1), \text{ with } t_2 \text{ very close to } t_1$$

As I wrote in the introduction, I'm really sorry for the equations. I know many tend to run away from them. Rest assured, you won't fail to grasp or appreciate the point of this book if you can hardly look at them without trembling. My point is just this: encoded in these two lines, you have the motions of the planets, of the moon, of every tennis ball, soccer ball, and, well, *every* ball. You have the tides in the ocean and many more phenomena. If you add to these lines the few additional equations you'll find in the next pages, you explain how TV works, radio, computers, the Internet, cars, rockets—every single device there is.

The first equation states that if I have two masses, M_1 and M_2, there is a force pulling each towards the other, with an intensity proportional to the product of the masses and inversely proportional to the square of the distance $r_{1,2}$ between them. The coefficient of proportionality G is called the gravitational constant. Now that I know the force acting on my object, I need to know how it's going to move. That is the role of the second equation. This second equation tells me that if at time t_1 the velocity of an object with mass M is $v(t_1)$, and if a force F acts on this object, then its velocity at a time t_2 very close to t_1 is $v(t_2)$, as given by the formula.

Things are mathematically more technical, but one can easily guess how these equations allow you to predict the trajectory of any body. If you know where your planet is now, you can compute the force the sun exerts on it. Then, if you also know the velocity of your planet right now, the second equation tells what its velocity will be in a short time. You can thus deduce where your planet will be in a short time, and you start all over again to guess the trajectory at later times.

The important point here is that these equations allowed Newton to derive every single law Kepler had devised. Just from these two lines, Newton could predict the orbit of every known planet. He could also predict the orbit of the moon and of Jupiter's satellites. But more than that: he could predict the trajectory of an apple (falling on his head?), as well as explain the tides. Suddenly, apparently disconnected phenomena such as the motion of the earth around the sun, the fall of an apple, or the tides in the Atlantic Ocean were found to derive from a single, and indeed quite short, principle.

Were Kepler's laws proved "wrong"? Not really. Kepler's laws were derived from observations, measurements. If there was agreement between these laws and observations *before* Newton, the agreement is obviously maintained *after* Newton. Newton's law could replace Kepler's because it had unifying and predictive capabilities that the latter did not have (Kepler didn't say anything about tides and falling apples).

The successes of Newton's theory were such that important discoveries could be made *trusting* it. Here is how: at the end of the eighteenth century, the French mathematician Laplace[1] computed accurately the orbit

---

1. The same Laplace who didn't need the hypothesis of God.

of the planet Uranus. In the first half of the nineteenth century, astronomers observed slight differences between the calculated motion and the real one. Was Newton wrong? To start with, they tried to solve the problem assuming Newton's laws were right. They found that *if* Newton was right, the discrepancies observed in the motion of Uranus could be explained assuming an unknown planet was following an orbit they calculated. They thus told the astronomer Johann Gottfried Galle something like, "Look in this direction, at that moment. You should see a planet with such and such properties." On September 23, 1846, Galle looked where he had been told to look and found the expected planet, which was called Neptune. That pattern has repeated over and over in the history of physics. Planets and particles have been discovered trusting known laws, and reasoning that "if this law is right, then this planet, or this particle, must exist" (see table 1, page 34).

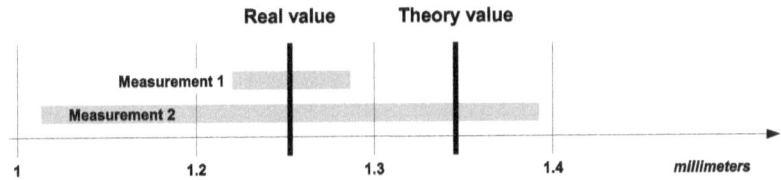

Figure 3: Suppose you want to measure something between 1 and 1.4 millimeters long. The measurement technique you choose has uncertainties (gray bars). For technique 2, your theory is "right." For 1, it is "wrong."

Newton's laws have been considered valid as long as they could explain observations. Now, it is obvious that you cannot measure the position of a planet with a millimeter precision. As shown in figure 3, a given theory cannot be proved wrong if the measurement uncertainties can't discriminate its predictions from reality. We thus need to emphasize that Newton's laws were considered valid as long as they matched observations *up to measurement uncertainties*. When the first discrepancies appeared between Newton's laws calculated consequences and observations, as in the case of Uranus, they could be explained within the very framework of these laws.

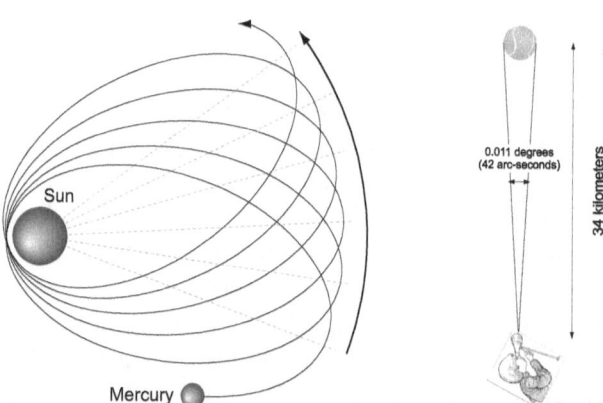

Figure 4: *Left:* The orbit of Mercury around the sun. Mercury follows an ellipse whose axis, the dashed line, slowly rotates with time. The sketch is not to scale.
Right: Newton's laws miscalculate the rotation of Mercury's orbit by 0.011 degree per century. How big is 0.011 degree? This is the angle under which you'd see a tennis ball 34 kilometers away.

Then came the Mercury issue. By the second half of the nineteenth century, astronomers found its motion differed slightly from the prediction of Newtonian mechanics. The ellipse it follows around the sun does not close perfectly, as if it were slowly turning around the sun, as pictured in figure 4. Accounting for every possible effect,[2] Newtonian mechanics predicts the axis of the ellipse does turn, but by 5,557 arc seconds per century. One arc second is only 1/3,600 of a degree. Newton, in other words, thus states that Mercury's ellipse turns by 1.54 degrees every century. The problem is that the measured value is 5,599 arc seconds per century. Note that the discrepancy is not that big: 5,557 versus 5,599 (0.75 percent). We miss 42 arc seconds per century—only 0.011 degree every 100 years! Figure 4 illustrates how *small* this angle is. It is the angle under which you see a tennis ball 34 kilometers away. Yet—and this is part of how science works—if the discrepancy cannot be explained within the framework of the known laws, something is "wrong."

Encouraged by the Neptune success, some assumed that a yet to be found planet was disturbing Mercury's orbit. They computed the orbit it ought to have to produce the missing 42 arc seconds per century, and called it "Vulcan" even before it was observed. In 1860, the French astronomer Urbain Le Verrier began the hunt for Vulcan. He died in 1877 without having

---

2. Mainly the influence of the other planets.

found anything. Others kept searching, but eventually gave up. Vulcan was never found. The missing planet hypothesis had failed for Mercury.

If, after having evaluated every possible effect, the theory is unable to explain observations, something else must be accounted for. The known law must be revised. Newton's law is still valid for a huge number of phenomena. This Mercury finding is not going to "undo" those agreements. Indeed, if we could measure the orbit of any planet with enough precision, we would find Newton's law does not fully explain what is observed. Einstein's so-called general relativity is needed here (more on this below). For the other planets of the solar system, general relativistic effects are much smaller than for Mercury [23]. They have been measured for Venus and the earth and found in agreement with Einstein's theory [10].

We therefore come to a very interesting conclusion: Newton's laws were never absolutely "right" nor absolutely "wrong." They simply predict gravitational motion up to a certain precision. If you're interested in the motion of the planets beyond Mercury, or in the fall of an apple, or in the orbit of the moon or Jupiter's satellite, the precision of Newton's laws beats observational uncertainties. But for Mercury's orbit, the departure from Newtonian mechanics is large enough to be observed. Still, Newton's accuracy remains outstanding: he is only wrong by the angle you see a tennis ball from 34 kilometers—per century (figure 4)! Call it a *big* mistake?

So what is there beyond Newton? The missing 42 arc seconds per century remained a mystery until, in 1916, Einstein came up with the theory of general relativity. Einstein applied his theory to the Mercury problem and found the missing arc seconds. To this day, Einstein's theory has never been found wanting, in spite of some truly weird predictions! For example, atomic clocks in a plane were found to turn at a *different* rate than those left on the ground, and by the predicted amount of time [46]. Why? Because as Einstein's theory predicts, time goes faster as you move away from Earth.[3]

---

3. This time shift with altitude, among others relativistic effects, is verified every day by any owner of a Global Positioning System, the famous GPS. The clock of a GPS satellite loses 0.00005 second per day with respect to another one left on the ground [91]. The position of the GPS is eventually determined measuring the distance between the apparatus and at least 4 satellites. These distances are in turn measured looking at the time it takes for the signal to go from the satellites to the receiver. Given the value of the speed of light, a time shift of 0.00005 second a day yields an error of 1.5 km a day. The few meters precision required for the GPS to tell you which road you're driving would be impossible to achieve without Einstein's theory. The overall GPS system is a 24/7 experiment constantly confirming its validity.

Validity Domains

So is Einstein "right"? We would rather say that Einstein's theory keeps predicting every observation made so far with an accuracy smaller than the measurements error bars.

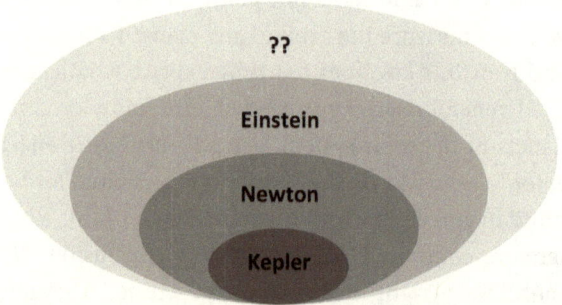

Figure 5: When a theory in physics is revised, it is necessarily contained within its successor. Newton's theory implies Kepler's, as Einstein's implies Newton's. Observations explained by Newton remain valid even when Einstein's theory appears, because the new theory cannot cancel previous experimental agreements. Nobody knows what is beyond Einstein.

Has Einstein's theory got the last word? No. General relativity is expected to fail in situations when the so-called quantum effects should play a role. I will say more about these "quantum effects" later. They basically have to do with how particles behave on very small scales, like the electrons in an atom. It has so far been impossible to do an experiment in which both gravitation and quantum effects come into play simultaneously. If, or when, this happens, Einstein's theory should be found wanting, and hopefully, experimental data will help build a new theory. Figure 5 schematically represents the situation. This new theory will be to Einstein's what Einstein's was to Newton's: it will contain it, expand it, but will not cancel all the experiments in which Einstein was found right.

Indeed, the hierarchy of theories highlighted in figure 5 illustrates the important concept of *validity domain* for a theory. Einstein explains everything up to measurement uncertainties in the solar system, and Newton, everywhere but for Mercury (roughly).

SOME MISCONCEPTIONS ABOUT SCIENCE   27

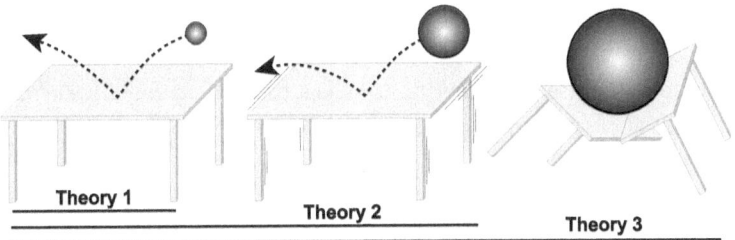

Figure 6: Theory 1 is valid as long as the ball does not shake the table too much; theory 2, as long as it does not break it; theory 3, always valid.

An illustration should help us visualize the idea of validity domain. Suppose you want to elaborate theories describing what happens when you drop a ball on a table. You may consider three kinds of balls, as shown in figure 6. The small ball just bounces, which renders the calculations quite easy. This will be theory 1. But if the ball is heavier, the impact could shake the table. As a result, the bounce may be affected, and the calculations will be a little more complicated. This will be theory 2, which accounts for such possible effects. Finally, you may want to account for all possible masses for the ball, up to a very big one. In such a case, the ball could be so heavy that it simply crushes the table. This will be theory 3, which incorporates every possible effect.

While it is clear that complexity rises from 1 to 3, we see that each one has its validity domain. Theory 1 is valid as long as the ball does not shake the table too much; theory 2, as long as it does not break it; theory 3, always valid.

Suppose someone says, "Okay, theory 1 predicts the ball bounces in certain ways. But science progresses, and sooner or later, your theory 1 will be outdated." The mistake is obvious. Granted, theory 1 is limited. If you do an experiment with a heavier ball, it won't predict the right bounce. But once you have found a good theory for heavy balls, it will necessarily merge with theory 1 for light balls. We can thus say that theory 1 is *definitively right, within its validity domain.*

Still, someone (probably the same person) could claim, "Okay. I understand that progress cannot cancel the validity of theory 1 with light balls. I understand theories developed for heavier balls will have to merge with theory 1, when applied to light balls. Yet, couldn't theory 1 be wrong from time to time, even for light balls well within its validity domain?" Indeed, that would be the definition of a *miracle*. Would you expect a ping-pong ball to break the table? Would you expect a pea to shake it? Indeed, breaking the table this way would be just that, a *miracle*, an "event that is not explicable by natural or scientific laws," as the *Oxford English Dictionary* states. I don't claim they don't exist. I just claim natural laws don't change from one day to another.[4]

It is important to realize that you can't simply define a miracle as "something strange." If a one-year-old baby sees you flying, will he find it "strange" and start screaming? Probably *not*. He doesn't know people can't fly. There is more: would you consider passing through a wall without injury a "miracle"? I guess so, yes. Indeed, it would be so. Still, if you were a little particle, like an electron or a proton, the laws of quantum mechanics (see Appendix A) would tell you it is *possible*. This is called the quantum "tunnel effect." For objects our size, the probability is incredibly small. But for elementary particles, it works very well. Some physicists even built a microscope using this principle,[5] for which they were awarded the 1986 Nobel Prize in physics. So both babies and tunnel effects demonstrate that the *Oxford English Dictionary* is right. There are miracles that wouldn't surprise a baby, and routine events that seem miraculous.

Simply put, successive theories with ever-increasing validity domains have to contain predecessors that were found valid within more restricted validity domains. This domain can have to do with the proximity to the sun (gravitation), the weight of a ball, or the size of the system (quantum mechanics), but the pattern is always the same.

Let me just end this section by quoting from a book that Einstein wrote with Leopold Infeld, titled *The Evolution of Physics* ([31], page 152):

---

4. This book explains how astronomical observations back up this claim. But there is *another* backup. It is quite technical, which is why I have chosen to park it in this footnote. Most laws of physics are indeed *conservation* laws. Many physical quantities are conserved as the world evolves. Energy is just one of them. There is a mathematical theorem, called "Noether's theorem," which states that every conservation law is related to a *symmetry* of the laws of physics. Symmetry? The laws of physics shouldn't change whether I study them here or in the room next door. This is the symmetry of *translation*, which corresponds (through the theorem) to the conservation of "momentum." Nor should they change whether I do my experiment looking at the window or at the corridor. That is symmetry of *rotation*, paired up with the conservation of "angular momentum." Now, you may have heard of the conservation of *energy*? Which symmetry is associated with this one? Translation in *time*. As weird as it seems, stating that energy is conserved is equivalent to stating that an experiment performed today will give the same result tomorrow. If the laws of physics could change with time, then energy would *not* be conserved and our industrial civilization would be complete chaos.

5. You can *see* individual atoms with it!

We could say that creating a new theory is not like destroying an old barn and erecting a skyscraper in its place. It is rather like climbing a mountain, gaining new and wider views, discovering new and wider connections between our starting point and its rich environment. But the point from which we started out still exists and can be seen, although it appears smaller and forms a tiny part of our broad view gained by the mastery of the obstacles on our adventurous way up.

Dating Relies on Known Physics

Here is a very important point for dating techniques. We've just seen that every law has a validity domain where it can be trusted. Within this validity domain, a departure from the law is a miracle. The question then arises: when the known laws of physics state that the world is older than six thousand years, do they speak from inside the validity domain, or outside? In other words, do we need to use these laws in a domain where they are not expected to work? Definitely *not*.

Chapter 4 will deal with the age of the universe derived when observing remote stars. The laws involved here are just good old standard physics. The equations describing light propagation[6] are almost 150 years old. The equations we use to observe the chemical composition of these stars are almost 100 years old. Physicists know too well that we are missing some laws. But the day we find them, if ever, they will have to reproduce precisely what today's laws tell us about these observations.

Then, chapter 5 addresses what astrophysics teaches on the so-called radioactive decay rates. Their stability with time is the key to their dating capabilities. These rates depend on the laws of nuclear physics. And the laws we know don't have any problem computing them. When we observe some remote nuclear processes in the universe, what we see lies completely within the validity domain of these laws.

Whether we deal with atomic physics (chapter 4) or nuclear physics (chapter 5), the key observations do *not* involve anything outside the validity domains of the equations we have. As explained in chapter 4, observations imply these laws were the same in the past. They also imply that any law yet to be found will have to teach us the same lessons.

---

6. The Maxwell's equations. See equations (2), page 53.

You Cannot Build the Theory You Want

How is it that nature has laws? How do you find them? How do you build a theory? An answer, probably partial, to these questions lies in the finding that the world is logical. So if you want your theory to be consistent with both *logic* and experiment, you don't have much choice. These two guards are closely monitoring your efforts.

Any theory must therefore meet two stringent requirements: it must be consistent with logic, and it must be consistent with experiment. Because the former requirement is less obvious than the latter, I will start by emphasizing the logical, or mathematical, consistency issue.

To prove that 2 + 2 = 5 is wrong does not require any observation on my part. I just have to imagine: if I place 2 apples on a table, and then add 2 more, there will be 4 apples on the table, not 5. Math is just the extension of this kind of logic. That 2 + 2 = 4 is a truth intuitively obvious. By elaborating on such obvious truths to produce new ones, and elaborating again on these new ones, math generates truths far less intuitive, but still as true as 2 + 2 = 4. Let me provide two examples:

- It is *impossible* to find three numbers x, y, z all larger than 1, fulfilling $x^n + y^n = z^n$, if n is a number larger than 3. By "number" here I mean one without a decimal point, like 3 or 4, and *unlike* 3.1 or 4.5. In math, such numbers are called *integers*. You can choose any value for integers n, x, y, and z, yet the equality will *never* be fulfilled.[7] Note that you can't prove this by trying every possible combination. There is literally no limit to their number. You could program the most powerful computer on earth to try every possible number, let it run 100 years, and still fall incredibly short of what would be left to check. But pure logic can do the job.[8]

- The sum of all the inverse squares,

$$1 + \frac{1}{2^2} + \frac{1}{3^2} + \frac{1}{4^2} + \frac{1}{5^2} + \ldots = 1 + \frac{1}{4} + \frac{1}{9} + \frac{1}{16} + \frac{1}{25} + \ldots$$

---

7. The symbol x^n means the integer x is multiplied n times by itself, that is, $x^n = x*x*x* \ldots *x$, where x is written n times. This identity has a solution for $n = 2$ because $3^2 + 4^2 = 5^2$.

8. This result was set forth without a proof by the French mathematician Pierre de Fermat in 1637. It took 358 years to find a proof. That is to say, I can't explain here *how* logic does the job (indeed, I really can't, for I don't understand the proof myself).

is *exactly* $\pi^2/6$. Take a computer and start evaluating the sum. After having summed up to $1/5^2 = 1/25$, add $1/6^2 = 1/36$, then $1/7^2$ and so on. You will observe that as you keep summing, the result gets closer and closer to $\pi^2/6 = 1.64493\ldots$ Math can prove this, that is, show why this is so.[9] Here again, no computer on earth can *prove* this equality. Why? Because the number of digits after the decimal point is literally infinite. Even if a giant computer could give you billions and billions of these digits, you wouldn't even come close to the real number. Only math can tell you that the infinite sum of all the inverse squares *is* $\pi^2/6$.

Because math is just an extension of the 2 + 2 = 4 type of logic, a theory must be mathematically consistent if it is to describe the real world. This kind of consistency restricts greatly the number of ways you can modify a theory that has failed to predict an observation.

Math is a powerful tool. Take a starting point, translate it to math, and you can derive a virtually infinite number of quantitative *consequences* of your starting point. Take the flow of water in a pipe. Do you agree that what enters the pipe *must* come out, though you may have a "traffic jam" inside? The mathematical translation of this sentence is (again, sorry for the equation):

$$\frac{\partial \rho}{\partial t} + \nabla(\rho \mathbf{v}) = 0$$

Believe me. This is *really* like translating to Chinese or German. ρ is the density of water at any given point of the pipe, and $\mathbf{v}$ its velocity. If you agree that "what enters the pipe *must* come out, though . . ." then the equation above can tell you *all* the consequences of this principle. What if it fails to successfully predict the outcome of an experiment? The problem has to lie in the principle, not in the math. Don't kill the messenger. But if you found the right basic principles, a few equations will take you a long way. That's how the millions of processes we understand eventually come down to a few basic principles.

As stated before, any new theory must obviously be consistent with observations—with new observations that previous theories couldn't explain, of course, but also with *past* observations that previous theories *did* explain correctly. What would you think of a theory that explains only

---

9. See Wikipedia, "Basel problem."

the last mysterious experiment, but fails on everything else? So your new theory must be mathematically consistent with itself *and* with the old ones.

As you climb the theory hierarchy levels, the constraints multiply. For example, Newton's formula (1) has gravitation varying as the inverse *square* of the distance. But it could be different without breaking the internal consistency of the theory. Gravitation could decrease with distance d, but with the *cube* of it ($d*d*d$) instead of the square, so that going twice as far away from the sun reduces its gravity by a factor of 8 instead of 4 ($2*2*2$ instead of $2*2$). Why the square and not the cube? To this question, Newton answered that the square law was required to fit Kepler's law. But Einstein *proved* this is a square. In Newton's theory, the square is a number elected to fit observations. In Einstein's theory, the square itself is a fruit of the theory. For Newton, the square is a free parameter. If we had a cube, Newton would just have set a "3" here. But if nature had chosen a cube, Einstein would be wrong, because it demands a square to be logically consistent.

Figure 7: *Left*: Claude Monet's *Garden at Giverny* (c. 1922). *Right*: Leonardo da Vinci's *Mona Lisa* (c. 1503). The missing piece of the puzzle must have the correct shape *and* the correct content.

We find here a key feature of this hierarchy of theories. The more you climb the theoretical ladder, the "stiffer" the theories get. Newton could adapt to a different law of gravitation. Not Einstein. The theory beyond Einstein will probably be even stiffer, for it will have to encompass Einstein's gravity *plus* quantum mechanics. It's pretty much like searching for the missing piece of a puzzle. Suppose you need one piece to complete Claude Monet's *Garden at Giverny* (see figure 7). The missing piece must fit with the neighboring ones; its shape must be identical to the hole left by the others. In our theory world, we'd say that the ultimate theory must join the other ones in their respective domains of validity. In other words, it must also have apples falling and atoms forming molecules. But that's not enough. You can't just take the right shape from Leonardo da Vinci's *Mona Lisa* and plug it into Monet's *Garden at Giverny*. The piece wouldn't match the painting. Again in our theory world, that would represent the issue of internal consistency. Bridging with known theories and experimental results is not enough. It must be done in a mathematically consistent way.

So, how do you modify Newton to obtain Mercury's orbit? You could imagine changing some constants in the formulas—G, for example—to fit the observed orbit. But then you lose agreement on the orbits of the other planets, and you cannot end up with a theory that needs a different G for every planet. (Well, you can, but it is completely inconsistent and can only be a temporary patch until you find a logically consistent way to modify the laws.)

How did Einstein modify these laws? By rethinking the basic premises upon which they rest—by noting, for example, that Newton's formulas imply an instantaneous action. If the sun loses mass *now*, the formula tells me that the intensity of the force with which it pulls the earth changes *now*. Indeed, Einstein revisited almost all of physics through the so-called special and general relativity theories. These two theories form a consistent "block," predicting that gravity travels at the speed of light, so that if the sun loses mass *now*, it will pull the earth a little less only *later* (see reference [55] for a complete, rigorous, and horribly mathematical exposition).

As emphasized previously, so far Einstein's theory has been incredibly successful at predicting all sorts of effects. But the point of this section is that its success comes from its mathematical consistency. The consistency requirement is a powerful guide when it comes to elaborating a theory, simply because the world is logical. More intriguing, this requirement has repeatedly allowed for predicting the very existence of entities. We're no longer talking here about predicting how matter behaves. We're talking about predicting the existence of yet unobserved forms of matter. Table 1 displays a list of particles (plus the planet Neptune) whose existence was logically inferred *before* they were observed.[10]

---

10. See Wikipedia, "Timeline of particle discoveries."

Name	Predicted	Observed
Neptune	1846	1846
Antielectron (Positron)	1931	1932
π Meson (lies inside the atomic nucleus)	1935	1947
Quarks (building blocks of protons, neutrons)	1964	1969
W Meson (lies inside the atomic nucleus)	1968	1983
Z Meson (lies inside the atomic nucleus)	1968	1983
Higgs boson	1964	2012

Table 1: List of physical entities whose very existence was theoretically predicted before they were discovered.

When it comes to building a new theory, there are rules to follow. And these rules are not made by humans. Scientists are not responsible for the fact that 2 + 2 = 4, nor are they responsible for the result of their observations. Without logical and observational consistency, a theory can only be a chimera with little, if any, connection to the real world.

Contrary to the claims of some postmodern philosophers, science is not merely a social construction. The planes they take to attend their congresses do not fly because scientists were lucky enough to find a theory that pleased them *and* that turned out to be okay with planes. Computers don't work because scientists have decided electrons in metal go in such and such ways. Electrons in metal have always behaved the same way, whether there were people to understand them or not.

Simply put, there is something out there to understand, and it does not care about whether you will like it or not. All you can do to chase it is trust logic and submit to observation.

Let us now deal with a series of misconceptions regarding the scientific mindset.

Physicists Don't Live to Prove the Bible Wrong

The 7.5 billion euro Large Hadron Collider (LHC) near Geneva was built to see whether a particle called the "Higgs boson" exists (it does). *Extremely* unfortunately, journalists love to call it the "God particle." As a result, many believers are upset and feel that scientists work just to get rid of God. I read the following sentence about the LHC on a Christian website: "This is a totally secular effort by scientists to explain the universe without a Creator."

So let me be clear: physicists don't enter their laboratories every morning wondering what they will do today to undermine the authority of the Bible. Whether or not they believe is absolutely not an issue when they're doing science. Some are Christian, others Muslim, still others agnostic or atheist. It may be (it's probable, even) that when thinking back about their work, their belief system comes into play. But physicists don't choose to study such and such a topic—and don't choose to come to such and such a conclusion—because of their religious beliefs.

The missions of the Hubble Space Telescope, the Planck Space Observatory, and the Chandra X-ray Observatory have absolutely *nothing* to do with gathering enough data to prove the Bible right or wrong.

At a more individual level, we don't choose our research topics, nor do we choose what we are going to find, based on our spirituality. We can decide to tackle such and such research question for many reasons: Am I interested in it? Is society interested? Will it be useful to someone now or in the future? Will I be able to get funding easily? Is it too difficult a question for me? Might I get a better job if I solve this problem? Might I become famous, even? None of these has anything to do with whether or not our results will be in harmony with the Bible. At any rate, one thing is sure: proving the Bible wrong is not among our top twenty daily concerns. I hope I'm not being too blunt here, but frankly, while in the lab, we just don't care about proving or disproving the Bible.

Physicists Don't Overprotect Their Theories

Still, contrary to popular perception, physicists don't act like Latin mothers (or Nemo's father) with their theories. They don't overprotect them. Instead, they do everything they can to test them, push them to the limit, and see how they endure the stress. For clarity, let me explore an example.

The "standard model" of particle physics is the theory built on quantum mechanics that describes the behavior of the world at the microscopic level. By smashing atoms against each other in particle accelerators, a whole zoo of particles has been discovered beyond the well-known electrons, protons, and neutrons. How do they interact? How do protons and neutrons hold together in the nucleus of an atom? These are questions answered by the standard model.

The largest particle accelerator ever built, the aforementioned Large Hadron[11] Collider (LHC), began to operate in 2008. Its goal was to put the standard model to the test and also to confirm it within the limits of its validity domain by finding the Higgs boson. In July 2011, the scientific journal *Nature* reported a work session at LHC where experimental results were contrasted against the standard model theory [17]. As the article puts it, "observations matched the predictions so perfectly that many of the numbers were identical." But instead of satisfaction, the journal reports that scientists showed a "hint of disappointment." Why? Because as long as you can predict what you observe, you don't learn anything new.

A year later, on July 4, 2012, the LHC people announced that they had found what they were looking for: the famous Higgs boson. Yet, less than two weeks later, the journal *Science* wrote, "physicists seem to hope the new particle won't quite match the mug shot of the standard model Higgs boson" [22]. At the very moment the "Higgs" was found, these guys started to expect it would not be exactly as predicted! How is that? René Thom, a French mathematician,[12] said [92], "When you know where you go, you don't go very far." Science progresses only with the unexpected, and *truly* plays by this rule.

The LHC has been built to create conditions where the standard model could *fail*. In 2008, the very year in which the LHC started to operate, *Nature* published an article [16] about it titled "The Race to Break the standard model." Again in *Nature*, a physicist declared in 2010, "From a mathematical perspective, we know that the standard model must be wrong" [64].

I could add quote upon quote, in every major field of physics. These statements are not exceptional at all. From what we've seen before, you now understand that "the standard model must be wrong" means "the standard model has a limited validity domain, and the LHC should allow us to probe what happens outside it." The "hint of disappointment" comes precisely from this: as they venture *beyond* the validity domain of what is known, people expect to be surprised. Truly, scientists do not fear their theory could be proved wrong. Rather, they *hope* for it.

---

11. Protons and neutrons are two examples of hadrons.

12. I here confess my French bias, for similar quotes have been attributed to Oliver Cromwell and Christopher Columbus.

How the Scientific Literature System Works

This section deals with an absence of conception rather than a "misconception." Nevertheless, I think it will be useful at the present stage.

To start with, let me give you four test sentences: "Extreme solar flares can release up to 1,000 Joules"; "Data show carbon dioxide level is currently 2,000 parts per million in the atmosphere"; "Hydrogen could be a solution to the depletion of primary energy sources"; "Landau proved any small perturbation to a Maxwell-Vlasov equilibrium is not damped." A few comments about them:

1. They all look scientific.
2. The first three were found in *major* newspapers.
3. They all are completely *wrong*.
4. It is not easy to tell if you don't know much about the topic.

I hope this won't be too great a shock to you, but when it comes to the few topics I'm reasonably knowledgeable about, I've found that mainstream media are far from being reliable sources of information. And I have no reason to think the same media suddenly become trustworthy when they turn to topics I'm rather ignorant of—the economy, for example.

Why is that? I think the problem is simply that journalists are seldom, if ever, experts on the science topics they write about. In addition, their writings are not systematically and carefully proofread by experts who could detect mistakes *before* they are printed. These two problems are precisely the ones "scientific literature" is designed to avoid. When scientists around the world who conduct research on some topic want to share their latest findings, they don't use TV, newspapers, or radio. They do it through the scientific literature system. Here is how it works.

Suppose I'm an astrophysicist who just found something interesting. I would like other astrophysicists to know about it. So I, the author, write an article explaining everything, and send it to the editor of a scientific journal. You may be familiar with *Nature* or *Science*.[13] They are the most famous ones. But there are many more, such as *Astrophysical Journal* or *Physical Review*. Indeed, various journals are available in every field of research.

---

13. See nature.com and sciencemag.org.

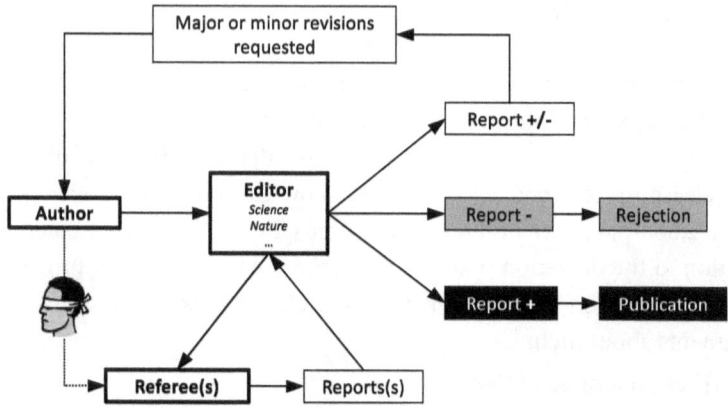

Figure 8: How peer-reviewed journals work. An author sends an article to the editor of a journal. The editor in turn sends the manuscript to experts in the author's field. Based on their reports, the editor may reject the article, accept it, or ask the author to make minor or major revisions. The author doesn't know who assessed his work.

Upon receiving my article, the editor sends it to one or more experts in my field of research who know, for example, that "Landau proved any small perturbation to a Maxwell-Vlasov equilibrium *is* damped." The experts, usually called "referees," write a report on my work and send it to the editor. Based on these reports, the editor may choose to reject my manuscript, accept it for publication, or ask me, the author, to make minor or major revisions. The reports usually answer the following questions for the editor:

- Is the paper free of logical mistakes? This has to do with the previous chapter. What you claim must be logically consistent.

- Is the paper consistent with experiments? If not, why? Any brand-new theory of gravity should predict that apples do fall. Now, if a new theory contradicts some more subtle experiment, the author is expected to know of it and to explain why. Of course, he'd better be convincing.

- Is the paper presenting something new? This requires a lot of experience. You may well have "discovered" warm water—something very useful, but something that millions of people already know of. In general, it takes a lot of time to know what has been done and what is left to do—to know the frontiers of knowledge in your field.[14] The editor

---

14. When you begin a PhD thesis, your PhD advisor gives you a topic to investigate. He acts like a mountain guide: a mountain guide knows the mountain and takes you directly to a good spot for skiing. Years later, when you too know the mountain, you will appreciate his advice. It takes years to know the scientific "mountains" of your research

of a good journal will do his best to *reject* research articles "discovering" something that has been known for ten years.

- Any additional criteria the editor is willing to set up. For example, *Nature* or *Science* require that articles be of interest to a broad audience. They want a biologist to be interested in the astrophysical papers they publish, and vice versa. In general, articles must report quite important scientific news.

As hinted by figure 8, my revised version may or may not be seen by the experts again. Such involvement of experts in the editorial process is the hallmark of peer-reviewed journals.

Two major points need emphasis here:

1. I don't know who my referees are. So even if I'm the godfather of my research community, and having me on your side is very good for your career, my referee can freely tell the editor that my paper is absurd, if indeed he thinks it is. Suppose the editor rejects my article because of this negative report; there is no way I can exact revenge on this poor referee. I simply don't know who he is.

2. You may wonder about the definition of "expert" in this world. How does the editor know that so-and-so is an expert to whom he can send articles on a given topic? Actually, it's quite simple. An expert in a field is someone who has already published on this topic in peer-reviewed journals. In other words, an expert is someone who has been acknowledged as such by other experts. It has nothing to do with having a PhD[15]—nothing to do with registering somewhere, passing an exam, or being on an Ivy League university's payroll. Just publish a few interesting papers on a given topic, and one day an editor will ask you to report on a manuscript someone else has submitted to his journal.

It may seem a circular definition, as the experts are the ones who decide who shall become experts. And, indeed, it is. But who else could decide? Who else knows that the claim "Landau proved any small perturbation to a Maxwell-Vlasov equilibrium is not damped" is 100 percent wrong, and why? The extent of our ignorance is so vast that only a tiny fraction of the overall scientific community can focus on a given topic. Who knows

---

field and to find the good research spots.

15. By the way, no one cares about your having a PhD or not when you submit a paper. The only requirement is to write something that meets the editorial criteria.

that a given measurement apparatus, used in a given experiment, is not reliable in such and such conditions? Who aside from the experts within the scientific community is better equipped to evaluate what is written on this topic?

Do these experts form an opaque private club systematically banning any controversial views? Are there politically incorrect truths they don't want to hear? Granted, the expert community within a field has a tendency to welcome some topics more than others. There are scientific "trends" and "fashions." But science is progressing nonetheless: your computer is working, doctors are healing people, DNA has been discovered, and planes are flying. That means consistency with logic and consistency with experiments are eventually the real guides. As Richard Feynman, already mentioned in this section, wrote in the 1986 report on the Challenger Space Shuttle tragedy, "For a successful technology, reality must take precedence over public relations, for nature cannot be fooled."

I'd now like to mention some interesting historical developments that, I think, definitely show that the scientific community welcomes being *deeply* disturbed.

In 1885, the French chemist Marcellin Berthelot wrote, "The world is now without mysteries." Yet, the scientific community has accepted since then that particles can apparently go through two different slits at once,[16] and that time goes by *faster* if you're not on the ground [46].

In September 2011, some physicists announced that they had observed particles called neutrinos going faster than light.[17] The impossibility of going faster than light lies at the center of Einstein's relativity and stems from the fact that mass increases without limit as you approach that speed, so that it takes an infinite amount of energy to accelerate something to such a velocity.[18] This mass increase is tested every day in particle accelerators and has never been proved wrong. Einstein published his article in 1905 (only twenty years after Berthelot's claim!), scientists accepted it, and it has successfully passed every single experimental test since then.

So what happened with these faster-than-light neutrinos? Did the scientific community ignore the experiment? Did peer-reviewed journals

---

16. See Wikipedia, "Double-slit experiment."

17. See Wikipedia, "Faster-than-light neutrino anomaly."

18. For those interested in a greater accuracy, Einstein' relativity does *not* forbid to go faster than light. It forbids to go faster if you were slower, and slower is you were faster. There are theories of faster-than-light particles, consistent with relativity [35].

refuse to publish anything about it? Absolutely not. Although the consequences would have been overwhelming, everyone agrees on this point: if an *observation* shows your theory to be wrong, you can't change anything but the theory. As it turned out, a mistake was found in the measurements, so these neutrinos were indeed *slower* than light [104]. But the anecdote is very instructive: I found sixteen articles on this discovery published in peer-reviewed journals from 2011 to May 2012, when I wrote these lines. Let me mention just three of them: two were published in the prestigious American journal *Physical Review Letters* [11, 26] and the third in the no less prestigious European journal *Nuclear Physics B* [43]. These two journals publish on a regular basis papers authored by Nobel Prize winners or prestigious scientists. Yet, they were not reluctant to print these neutrino articles. It took a while to find the error in the measurements, and in the meantime, the scientific community just played by its rules.

One last word about a free-access web database called "arXiv" (arxiv.org). Before, or even after, submitting their work for publication in a peer-reviewed journal, many authors, especially in astrophysics, post their papers there. There is *no* peer-review process with arXiv. It is therefore a very useful means by which to disseminate documents, but it lacks the legitimacy that comes with the peer-review process. Still, an anecdote involving arXiv should prove that the scientific community is not a stickler for its own rules: in 2002 and 2003, the Russian mathematician Grisha Perelman posted on arXiv a series of articles[19] claiming that he had proved the Poincaré conjecture. A conjecture is a mathematical claim that looks right, despite lacking any rigorous proof. This one was a century old. I can't tell you much more about this Poincaré conjecture because, frankly, I don't even understand the titles of these articles. All I know is that people who do understand found they were right, and they awarded Perelman the 2006 Fields Medal, which is the Nobel Prize for math. Although Perelman didn't play by the rules, it was obvious that he was right, and he wasn't denied the honor due to him.[20]

It is wrong to pretend that science is the product of the status quo among scientists who don't want to be disturbed. History shows scientists do allow themselves to be disturbed. I don't claim it never goes without resistance or inertia. The American historian Thomas Kuhn wrote a great book on this subject titled *The Structure of Scientific Revolutions*. But

19. The first one is here: http://arxiv.org/abs/math.DG/0211159.
20. Indeed, he *refused* the Fields Medal.

conspiracy theories about the truth being sequestered somewhere are simply nonsense. And yes, the peer-reviewed journal system has so far been found to be a reliable means of sorting out and disseminating genuine scientific advances. Just look around and see. Today's world, for better or worse, wouldn't be there without science.

The most trustworthy news about what's going on in science is there to be found, although not in the mainstream media. I sometimes cite Wikipedia in this book because I've checked the pages I refer to, because it is easily accessible, and, most importantly, because Wikipedia itself refers to primary, peer-reviewed sources. Otherwise, this section just explains the meaning of my bibliography section. I've tried to give you more than good books written by good journalists (there are some, really, like James Gleick) concerning the science I talk about. I've tried to connect you to the very "experts" who did that science in the first place.

So next time you hear about some unsung genius whose discoveries are supposedly threatening "established science," just check where he publishes his work. If he needs to invoke some conspiracy to explain away why he can't access mainstream peer-reviewed journals, something is wrong.

Math and the Real World

This paragraph is not mandatory for the overall comprehension of the book. It is here because it is short, and because it elaborates on the intimate relationship between math and nature. I love this topic!

You will have understood by now that you must know about math if you want to do physics. Still, there are many things you can understand without math. This is what popular science is all about. But if you want to start dealing with the numbers and make quantitative predictions, you have to use math. It's like driving: it is perfectly possible to *know* how to go from New York to Boston by car, without knowing how to drive. But if you really want to *do* it, you need to learn how to drive. In other words, you can know the route; you can even get detailed directions from one or more Internet sites. But if you really want to *go* there by car, you need to drive. In the same way, you can understand what Newton's laws are about. You can understand their validity domain and much more. But if you really want to use them to compute the earth's orbit, you need math.

Many have wondered about the role of math in physics. Some four centuries ago, Galileo wrote, "The laws of nature are written in the language of mathematics." In his 1936 essay "Physics and Reality," Albert Einstein noted that "the eternal mystery of the world is its comprehensibility." Later on, Eugene Wigner, winner of the 1963 Nobel Prize in physics, wrote a famous essay titled "The Unreasonable Effectiveness of Mathematics in the Natural Sciences."

Think about it: the equations we've met are expected to give correct *quantitative* answers. If they predicted some distance to be 2.5 millimeters but one's observations had 2.1 ± 0.1 millimeters, then something *must* be wrong, because the prediction falls outside the measurement uncertainty (see figure 3). Now, what could be wrong is that I couldn't solve my equations exactly, so that 2.5 millimeters was not the answer to the equations, but only an approximation of the answer. This happens frequently when it comes to analyzing complicated situations. For example, the equations determining the trajectory of your golf ball are well known. But computing the stopping point of your 250-meter drive with one millimeter precision is impossible. Not because the equations are wrong, but because to compute this, you must know about the wind, the air flow around the ball, the way it will bounce when hitting the ground—all quantities impossible to assess precisely enough.

But what if we conduct a "clean" experiment in which math can be done very precisely? Measuring Mercury's orbit was such a clean experiment because only gravity interferes, which makes math easy. Newton's answer was found to be wrong, and Einstein stepped in. We also know Newton's equations (1) page 21 give wrong answers for motions close to the speed of light or too close to a very massive object such as the sun. The so-called Schrödinger equation, which describes the world at the microscopic, atomic level (see Appendix A), also starts giving wrong answers for motions close to the speed of light.

We do have the equations replacing Newton's and Schrödinger's for large velocities. These are Einstein's general relativity and "quantum electrodynamics," respectively. So, what about them? In clean experiments in which it was possible to compute very precisely their predictions, they have so far never been proved wrong, regardless of the measurement uncertainties. One of the most incredible agreements between theory and experiment is the so-called anomalous magnetic moment of the electron, usually labeled g. This quantity determines how a single electron reacts to a magnetic field. It has been both measured [47] and computed [2] with extreme accuracy. The experimental and theoretical values g_{exp} and g_{teo} are:[21]

$$g_{exp}/2 = 1.001,159,652,180,73(28)$$

$$g_{teo}/2 = 1.001,159,652,18(6)(4)(2)(78)$$

The agreement between theory and experiment is measured by the relative accuracy

$$\frac{g_{exp} - g_{teo}}{g_{exp}} = 7.29 \times 10^{-13} = 0.000,000,000,000,729$$

That is, one part per 1,000 billion. In order to grasp such extreme accuracy, imagine you want to measure something as large as the distance from Los Angeles to Sydney (12,087 kilometers). The precision above amounts to predicting that distance with an accuracy of one-fifth of a human hair.

I didn't explain this to impress you (well, maybe a little). I did so to show that science really does go by the rules I explained. Devise a theory, test it to the limit, and if it fails the test, find a successor. Why are people measuring g to such precision? Mainly to test quantum electrodynamics. So stay tuned and wait for the next theoretical digit. If it fails to be a "0," there will be a scientific buzz.

---

21. The numbers between parentheses indicate uncertainties in a very specific way that there is no need to emphasize here.

4

More than Six Thousand Years Traveling

In a Nutshell

- The distance to some astronomical objects has been found to be *greater* than six thousand light years by a very simple method of measurement. (One light year is the distance that light covers in one year at the velocity it goes now.)
- If light has always been traveling at the same speed, it means it had to leave these objects more than six thousand years ago in order to reach us now. The universe is therefore older than this. Could it be that light went *faster* in the past?
- The speed of light is part of the known laws of nature. The laws we know don't change with time. Still, we could have missed something. Would a given experiment confirming these laws *now* give the same result if performed six thousand years ago? Indeed, it is *possible* to check.
- Every atom or molecule emits a very special kind of light, a kind of signature called "spectral lines." Theory predicts spectral lines observed *now* with extreme precision.
- The spectral lines we observe from remote objects in the universe are exactly *the same* as those we observe now in the laboratory, which means that atoms and molecules behaved in the same way in space, when they sent light to us. Hydrogen, oxygen, water, and so on had exactly the same properties when they sent their spectral lines to us as they do now.
- Observations confirm that the laws of nature were the same six thousand years ago (and more) as they are now. The speed of light was

45

therefore the same then, when light left these objects. This is not an assumption; it is an *observation*.

- Unless light was miraculously created in transit, it had to be up there more than six thousand years ago in order to reach us now.

There are many reasons why scientists have come to the conclusion that the universe is more than six thousand years old. The Antarctic ice sheet is nearly four thousand meters thick. Scientists have used it to analyze the change in atmospheric temperature, along with the concentration of methane and carbon dioxide, over the last eight hundred thousand years. The universe is therefore that old, at least. All three data series—namely, temperature and concentration of methane and carbon dioxide—are remarkably coherent with each other *and* with long-term cycles related to the earth's orbit around the sun [52, 15, 58, 60, 62]. The Greenland ice sheet provides records up to 250,000 years old [28]. Counting the rings of fossilized trees allows scientists to deduce that some trees were in existence more than eleven thousand years ago. This is one of the methods used to "calibrate" carbon-14 dating [86]. Astrophysicists observe objects far more distant than six thousand light years away. Again, one light year is the distance light covers in one year—nearly 10,000 billion kilometers. If we see objects that far away, then light had to travel that long to reach us. And radioactivity, this natural process changing atoms into other atoms, shows that some rocks must be billions of years old [102].

I won't review every single one of these evidences, for two reasons: first, I want to keep this book short, and second, I don't want to leave my area of expertise and so risk being inaccurate. Instead, I will focus on the last two. Astrophysics offers wonderful insights into the question of the age of the universe because here we can literally *observe* the past. We can geometrically *measure* the distance from astronomical objects, *observe* how the laws of nature were the same "over there" when light left them, and deduce that the light we see must have started its journey far more than six thousand years ago.

This point alone proves that the universe is older than six thousand years. Suppose your parents live 120 kilometers away from you. They just arrived, and you know their route to your home is all highway, with a speed limit of 120 kilometers per hour. You can deduce that they left home one hour ago, or so. Obviously, that implies that the universe was already there one hour ago. Now imagine that they live very, very far away—56,764,800,000,000,000 kilometers away. They just arrived, and you know their spaceship travels at 300,000 kilometers per second. You readily deduce that they left home six thousand years ago.[1] As a consequence, our universe already existed six thousand years ago. So it must be *older* than that. This is exactly what happens with starlight.

On a different timescale, our sun illustrates the same point. The sun-earth distance is such that it takes sunlight about eight minutes to come to us.[2] When you look at the sun now, you do *not* see how it is now. You see how it was eight minutes ago. This proves the world existed eight minutes ago. Granted, this is no big news, but things become more interesting when replacing eight minutes with ten thousand or thirty thousand years. Of course, you need to be sure your stars (your parents' home) are that far away, and that light, like your parents' spaceship, always traveled at the same speed.

As we shall see, the distance issue is solved by the reliability of the purely geometrical method used to measure it. Now, what if light was faster in the past? It could have left a star located ten thousand light years away at twice the velocity it has today, then progressively slowed down and reached us in less than six thousand years. As I will explain, the speed of light is part of the laws of physics. And observation tells us these laws were the same thousands of years ago, when light left these objects.

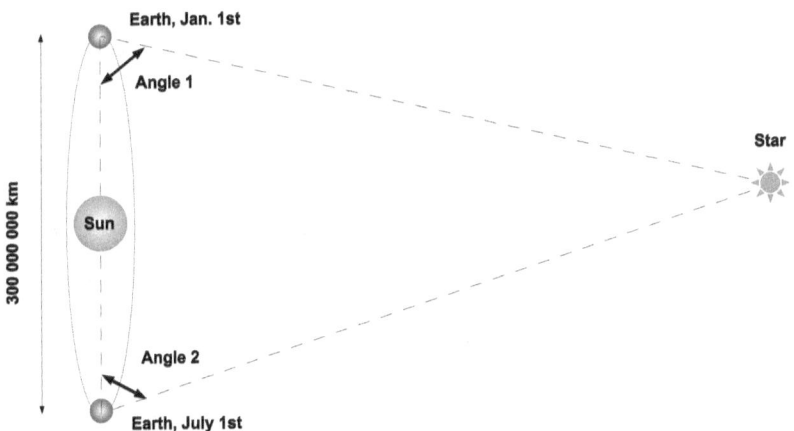

Figure 9: Knowing angles 1 and 2 and the diameter of the earth's orbit around the sun, we can determine the distance from the sun to a star.

---

1. Just divide 56,764,800,000,000,000 kilometers by 300,000 km/s. You'll find 189,216,000,000 seconds, which is precisely six thousand years.

2. Dividing 150,000,000 kilometers (the sun-earth distance; see figure 9) by 300,000 km/s yields 500 seconds, a little more than eight minutes.

Measuring Large Distances

The most basic method to measure the distance to an object is called the parallax method. It consists in looking at the object from one place, then moving a few meters and measuring how much I had to turn my head to keep watching it. This method can be adapted to measure the distance to remote stars, as represented in figure 9. Suppose we point a telescope toward a star on January 1 and measure the angle (angle 1). We wait six months to let our planet rotate half a revolution around the sun. Then, on July 1, we again point our telescope toward the same star and measure the angle (angle 2). One can imagine how, knowing the diameter of the earth's orbit, the two angles allow for the determination of the distance to the star. Because the earth's orbit is 300 million kilometers wide, it is possible to evaluate the distance to very remote stars.

Table 2 shows the measured distances to various astronomical objects. The distances are expressed in parsecs and light years. One parsec is the distance from which the earth-sun system is viewed under an angle of 1/3,600 degree. One parsec is 31 trillion (31,000 billion) kilometers. This unit is used here because it is usually the one used in the original articles. The measured distance is then divided by the velocity of light (nearly 300,000 km/s) and converted into light years. This other unit amounts to measured distances in terms of the distance light covers in one year. One parsec is 3.26 light years. Finally, the table also indicates the sources of the data, that is, the bibliographical references to the original reports written by the authors of those measurements. Note that because the number of cute names (and the imagination of scientists) is not infinite, astronomical objects usually receive quite opaque labels like "G9.62+0.20" or "IRAS 00420+5530."

Although the table contains twenty-six entries, more data are available. I just searched the word *parallax* on arxiv.org and found another article [73] whose authors measured two stars 10,900 and 10,500 light years away. They are *not* in the table. Soon, the European Space Agency will launch the "star surveyor" Gaia, which will measure by the parallax method the distance to stars up to 10,000 parsecs, or 32,600 light years, away. The number of objects whose astronomical distance has been measured geometrically beyond six thousand light years is thus bound to increase dramatically, as Gaia is scheduled to register up to one *billion* stars!

Table 2 does more than provide one single measurement of one star located just 6,001 light years away. We find here twenty-two objects more than seven thousand light years away. If you look at the original articles, you will see that these measurements, like all measurements, have "error bars."

	Object	Distance (pc)	Distance (ly)	Source
1	SfR G75.30+1.32	9,250	30,155	[80]
2	SfR G27.36-0.16	8,000	26,080	[100]
3	Pulsar B1541+09	7,200	23,472	[20]
4	SfR G23.44-0.18	5,880	19,168	[18]
5	Pulsar B2053+36	5,500	17,930	[20]
6	SfR W51	5,410	17,636	[82]
7	SfR G9.62+0.20	5,200	16,952	[81]
8	Pulsar B1933+16	5,200	16,952	[20]
9	SfR W 51 IRS2	5,100	16,626	[101]
10	SfR G23.01-0.41	4,590	14,963	[18]
11	Pulsar B2154+40	3,400	11,084	[20]
12	SfR G35.20-1.74	3,270	10,660	[103]
13	Binary star Scorpius X-1	2,800	9,128	[13]
14	SfR NGC 7538	2,650	8,639	[67]
15	Pulsar B0136+57	2,650	8,639	[20]
16	Pulsar PSR J1559-4438	2,600	8,476	[29]
17	Pulsar B1508+55	2,370	7,726	[21]
18	SfR G12.89+0.49	2,340	7,628	[100]
19	Pulsar B2020-28	2,300	7,498	[14]
20	SfR G35.20-0.74	2,190	7,139	[103]
21	SfR IRAS 00420+5530	2,170	7,074	[66]
22	SfR G59.7+0.1	2,160	7,041	[101]
23	SfR S252	2,000	6,846	[77]
24	SfR G15.03-0.68	1,980	6,454	[100]
25	Pulsar B0818-13	1,960	6,389	[20]
26	Cygnus X-1	1,860	6,063	[76]

Table 2: Measured distances greater than six thousand light years to some astronomical objects, by the geometrical parallax method. See the full references in the bibliography. "Pc" = parsec; "ly" = light years; and "SfR" = star-forming region.

These bars account for the limitations of the measurement technique. But even when considering the lower estimate for the distance to these objects, all but Cygnus X-1, the closest one, remain farther than six thousand light years away.[3]

3. The lower estimate for Cygnus X-1 is 5,700 light years. By the way, Cygnus X-1 contains a black hole.

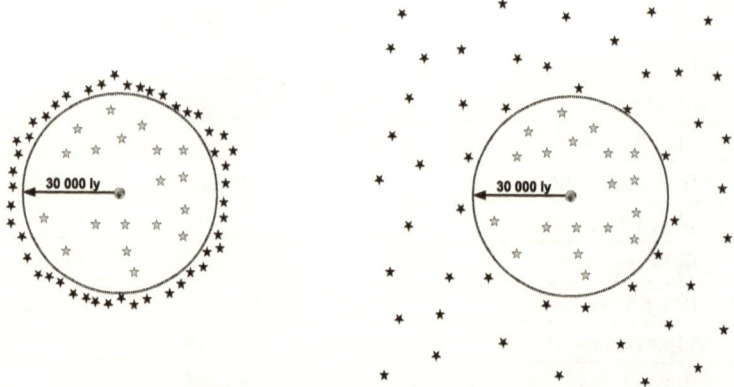

Figure 10: Two possible repartitions of the more than one hundred million galaxies farther than thirty thousand light years away.

There are many astronomical objects farther than six thousand light years. This is neither a theory nor an opinion; it is a fact. In fact, common sense tells us that millions of objects are much farther than table 2's champion at 30,000 light years. How do I know? I just consulted the NASA Extragalactic Database.[4] To this day, it contains 102,429,468 observed galaxies (and counting). The parallax method works for *none* of them, which means *all* are more than 30,000 light years away. Figure 10 offers two possible ways of placing some of these galaxies beyond the thirty-thousand-light-years limit. Who would choose the option on the left?

This threshold distance of six thousand light years is crucial for our problem. If we see these objects, it is because their light came to us in the first place. But light does not propagate instantly. If a star is located eight thousand light years away, then its light had to travel for eight thousand years to get to us. The star was already there eight thousand years ago, which implies that the universe is older than six thousand years.

Could a star created six thousand years ago, and located twenty thousand light years away, send us its light in *only* six thousand years? Could it be that its light traveled *faster* at the beginning? Suppose light was ten times faster than it is now during the first sixteen thousand light years. It would have taken only sixteen hundred years to cover the first sixteen thousand light years. Accounting for the four thousand years needed to cover the remaining four thousand light years, we find the complete journey would have lasted 1,600 + 4,000 = 5,600 years. Is that a possibility? This is discussed below.

4. See http://ned.ipac.caltech.edu.

Various Colors, Various Kinds of Light

Light does much more than reveal to us the distance to the object we are observing. Even to the naked eye, a blue car does not look like a red car. Light carries color according to what is called its frequency, or a related feature, its wavelength. Light is a wave made of particles we call photons (think about water molecules joining to form waves in the sea). The wavelength is the distance between two consecutive crests. Light also carries energy, as every photon carries some energy (solar cells exploit that). What is called the electromagnetic spectrum is displayed in figure 11. Light that we can see is indeed a very tiny portion of what is available. The wavelength of blue light is about 470 nanometers (it takes one billion nanometers to make one meter). Green is around 545 nanometers, and red around 680. Our eyes can see light between 380 and 780 nanometers. Below and beyond these visible wavelengths, there is light that we cannot see, such as the X-rays we all know from radiography, at a wavelength below ten nanometers, the microwaves of your oven, or the radio waves transmitting your favorite FM radio programs.

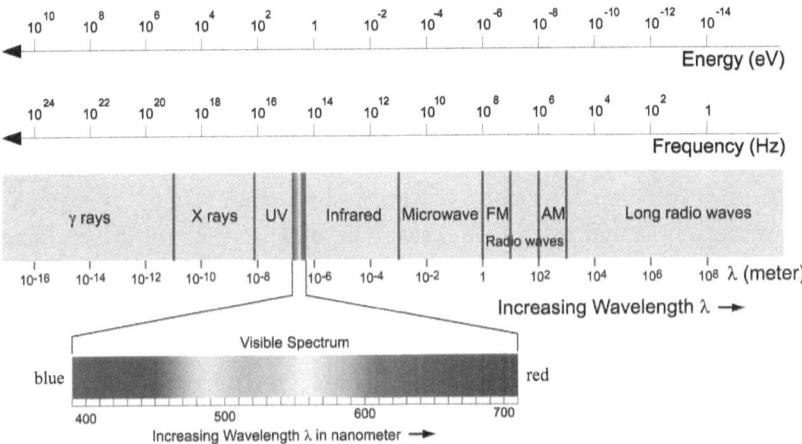

Figure 11: The electromagnetic spectrum. What we see is a very small part of what there is. Wavelength, frequency, and energy are interchangeable. The blue color is around 400 nanometers, and red around 700. Frequency is measured in hertz (Hz), which counts the number of beats per second. Energy comes in electron volts (eV), where 1 eV = $1.6*10^{-19}$ joules. One joule is the kinetic energy carried by a mass of 2 kg at a speed of 36 km/h. About the numbers: 10^2 is 100, 10^3 is 1,000, and so on; 10^{-2} is 0.01, 10^{-3} is 0.001, and so on.

When it comes to light, wavelength, frequency, and energy are interchangeable. Light at a given wavelength λ has the frequency f,

$$f = c / \lambda$$

where c is the speed of light. The frequency f is something everybody can relate to, as your radio tuner uses it. When you listen to a radio emitting at 90 MHz, your receiver is set to receive light with a frequency near 90 MHz. As you will have noticed, you can't *see* this light. Though your receiver *receives* something, you don't *see* anything entering it. But it is there, transmitting your favorite song. Light also carries an energy E given by

$$E = hf$$

where h is the so-called Planck's constant (see table 5, page 92). The bottom line is that wavelength/energy/frequency are various properties of the same kind of light. In what follows, I will sometimes refer to only one of them, depending on the problem we'll be discussing.

The sun sends us all kinds of wavelengths, and it is possible to separate them. The rainbow, for example, splits the sun's light and reveals its visible components. The parts of the sun's light that we *cannot* see, such as X-rays, can be detected and measured by instruments designed for that purpose. So when we look at an astronomical object, we can analyze its light and detect which wavelengths are being sent. What can we learn from this?

Matter is made of atoms. Atoms combine to form molecules, and molecules combine to form lots of things—us human beings, for example. Atoms, such as hydrogen, helium, carbon, and oxygen, are the building blocks of matter. Water, for example, is made of molecules consisting of two hydrogen atoms and one oxygen atom (the famous H_2O). Now, each kind of atom or molecule emits a very special kind of light. If you fill a tube with hydrogen, heat it, and analyze the light that comes out, you will observe only a number of definite frequencies, and nothing else. Out of all the possible frequencies displayed in figure 11, hydrogen emits *only* some. And apart from these, absolutely *nothing* else gets out of your tube.

Now, here is the important part: hydrogen has its preferred frequencies. So do helium, oxygen and every single kind of atom. So do molecules such as water, carbon dioxide, and so on. These frequencies are called spectral lines, or simply lines. When someone talks about "the hydrogen lines," or "a hydrogen line," he is referring to the complete set of hydrogen frequencies, or to any single one of them. Indeed, the spectrum can be used to *define* an atom. Because two spectra don't overlap, which means the spectral lines for two *different* atoms or molecules are *never* the same, you can define the atom by its lines. What if I look at a star and detect the hydrogen lines? It simply means there is hydrogen over there. And the same holds for helium, water, and so on.

The physical theory describing these lines is well known and *extremely* accurate. Those interested in more detail about it can jump to Appendix A. These laws of nature have been extensively tested, and the speed of light is part of them. What we are about to check now is that, whether we observe hydrogen lines (or others) on earth or in a star located six thousand, nine thousand, twenty thousand, or thirty thousand light years away, we observe the *same* lines. What does this mean? It means that the laws of nature describing them were the same when light left those stars. And since the velocity of light is part of these laws, *observations* tell us that light was traveling at the same speed six thousand, nine thousand, twenty thousand, or thirty thousand years ago. Let's now check all this.

Astronomy and Past Laws of Nature

The French author René Barjavel wrote a science fiction book titled *Ashes, Ashes* [6] (*Ravage* in French) that tells the story of a world where the laws of electromagnetism change from one day to the next. These laws describe how charges and currents create electric and magnetic fields. Every single electric device depends on them. The industry depends on them. Hospitals, traffic lights, computers, planes, radio, TV, the Internet, iPads, and iPods depend on them. These laws, called Maxwell's equations, can be written in the following mathematical form (please, don't run away!):[5]

$$\text{div } \mathbf{E} = 4\pi\rho, \qquad (2)$$

$$\text{div } \mathbf{B} = 0,$$

$$\text{rot } \mathbf{E} = -\frac{1}{c}\frac{\partial \mathbf{B}}{\partial t},$$

$$\text{rot } \mathbf{B} = \frac{4\pi}{c}\mathbf{J} + \frac{1}{c}\frac{\partial \mathbf{E}}{\partial t}$$

5. The electric currents and charges stand here for J and ρ respectively. If you set both to zero, which means you look at what's happening in a vacuum, these equations have solutions that we call *light*.

In Barjavel's novel, one day in the year 2052 electricity stops working as it always has. The equations above are modified—no one knows how—and the current **J** of the last equation stops flowing the way it used to—from the plug to the TV set. Complete chaos follows, and France, where the action takes place, quickly returns to the Stone Age.

So, is it possible that the speed of light varied in the past? Is it possible that the equations above changed? It is important to understand that changing the speed of light implies changing the laws of nature as we know them. Look at the equations above: c, the speed of light, is there, as it is in the equations of Appendix A. Changing c is not like having a different morning temperature or even another president. It is rather like deciding that the sun will not rise tomorrow. Or that the same tomorrow will have twenty hours instead of twenty-four. How confident are you that your plane will not crash? I know many people who are afraid of flying. But I've never met anyone afraid of flying because the laws of aerodynamics might change during his flight.

Indeed, many physicists have explored the possibility that the speed of light, together with other fundamental "constants," may in fact *not* be constant [94]. To this day, no convincing evidence for a variation has been found. Many scientists are involved in this field of research, as the variation of one of our "constants" could shed a light on the "beyond Einstein" theory.

The reasoning is simple: we know there must be something beyond Einstein, because we don't know which laws would apply in settings where both gravity and quantum mechanics play a role. The laws we know don't have any time variation included. So, any observation of such variation could provide clues about the laws we don't know.

You may have heard about doubts raised by the so-called fine structure constant. This number is formed after the quantities given in table 5 (see appendix A). It is usually denoted by the Greek letter α and reads:

$$\alpha = \frac{2\pi q_e^2}{hc} = \frac{1}{137.03\ldots}$$

The speed of light c is there. Could recent works on the variation of α help explain why light emitted twenty thousand light years away might arrive in less than six thousand years? Not really. Some of the most recent results in this respect have been published in *Physical Review Letters* [98].

The hypothetical (there is a debate) variation found is lower than 0.001 percent over some ten *billion* years. The bottom line for us: over the timescales shorter than one hundred thousand years we are interested in, no significant variation of any constant has ever been detected. There is absolutely *no* scientific debate about it.

Let us now focus on some objects listed in table 2, page 49, and confirm that the laws of nature were the same at least thirty thousand years ago. I'm going to pick three objects out of the twenty-six listed. For each one of them, I will show a few (only a few) observations evidencing that nature behaves the same way *there* as *here* in the lab. Indeed, I should write, "Nature behave*d* the same way there as *now* in the lab," since the light we see now was emitted in the past.

- The first one in the table, G75.30+1.32, will show water was behaving the same when light left it as today. This implies the laws of nature involved in water lines emission were the same, and so was the speed of light. Because this object is thirty thousand light years away, light left it thirty thousand years ago. The universe is thus at least that old.
- The second object I have picked is G23.44-0.18 (number 4), a little less than twenty thousand light years away. Here, some evidence for the constancy of the speed of light comes from the observation of ammonia and other molecules. G23.44-0.18 was already there twenty thousand years ago.
- Finally, I have picked Scorpius X-1, nine thousand light years away. Evidence in that case comes from hydrogen. Scorpius X-1 was already there nine thousand years ago.

It is impossible to go through every single observation. They are far too numerous. For every single object in table 2, many atomic and molecular lines have been observed. And they *all* fit what we see in the lab today. The evidences we're about to see are but a tiny fraction of the observational proofs that the laws of nature have not changed during the last thirty thousand years.

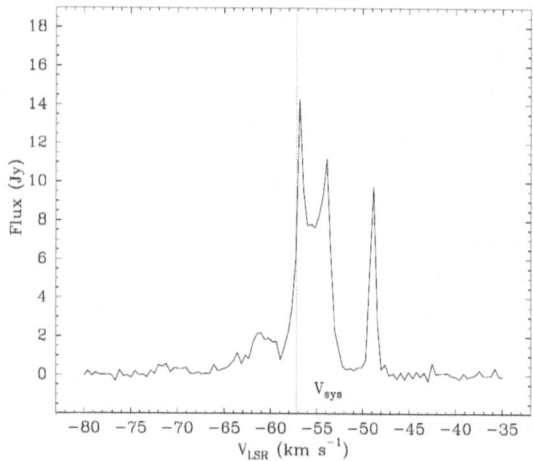

Figure 12: Water emission line at 22.2 GHz observed from G75.30+1.32, the first item listed in table 2, page 49. See the text for an explanation of the horizontal axis. From reference [80].

Thirty Thousand Light Years Away: G75.30+1.32

G75.30+1.32 emits, among others, spectral lines typical of the water molecule, at a frequency of 22.2 GHz. As already mentioned, molecules, like atoms, have their spectral signature. Each one of them emits its own kind of light. But the wavelength of light from molecules is some hundred thousand times *larger* than the one from atoms. The water emission line observed when looking at G75.30+1.32 [80] is represented in figure 12. And it is exactly where expected from the equations.

An explanation is here required regarding the *horizontal* axis. As you will see, we don't find nanometers, centimeters, or frequencies in GHz here. Instead, we find *velocities*, in kilometers per second (km s^{-1}). It turns out that what people plot here is not the frequency itself, but the velocity corresponding to the "Doppler-shifted frequency." This concept is fortunately *not* far from our everyday experience, as speed control radars on highways use this Doppler thing: as your car approaches it, the radar sends a pulse of light. Your car reflects part of the pulse, and the radar detects part of that reflected light. But because your car is moving, the reflected light has a slightly different frequency than the emitted light. By measuring that frequency shift, the radar can determine quite precisely your speed—and send you a ticket (or not). There is therefore a one-to-one correspondence between the frequency of the reflected light and your car's velocity.

Astrophysicists frequently use this effect to gain additional information from the spectral lines. Here, besides detecting water, we find the object flies away from us at 57 kilometers per second.

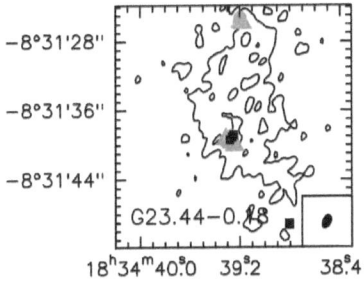 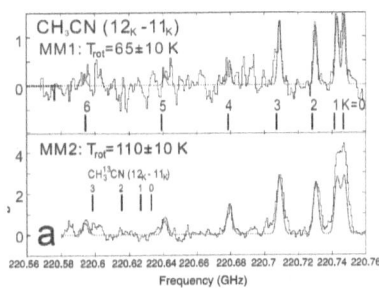

Figure 13: *Left*: Picture of G23.44-0.18, item number 4 in table 2, page 49, filtering an emission line of the ammonia molecule [24]. *Right*: acetonitrile (CH3CN) lines detected in each one of the two blobs forming the object [79].

Twenty Thousand Light Years Away: G23.44-0.18

We now turn to item number 4 on the list, G23.44-0.18. Here, a line produced by the ammonia molecule (NH3) has been detected, and a picture of the object taken, filtering this light only [24]. The result can be found in figure 13 (left), where the black contours represent places where the emission has a constant intensity (think about a topographic map). Furthermore, the squares, circles, and triangles on the picture mark the positions where water (H2O), methanol (CH3OH), and hydroxide (OH) lines, respectively, were detected by other authors [40, 96].

Also interesting is figure 13 (right) [79]. G23.44-0.18 is made of two main blobs. This picture shows the frequencies detected in each blob. These lines pertain to acetonitrile (CH3CN). The same article also contains pictures not shown here of the object filtering the light of carbon monoxide (CO), isocyanic acid (HNCO), sulfur monoxide (SO), and carbonyl sulfide (OCS). Note that the carbon monoxide involved here is made of one carbon and one "oxygen 18" atom. This "oxygen 18" is made of eight protons and ten neutrons instead of eight protons and eight neutrons for the most common form of oxygen. The spectral lines detection technique is so accurate that it reveals the kinds of atoms you find *inside* the molecules. In that case, $C^{18}O$ vibrations differ slightly from the usual $C^{16}O$ ones, just because $C^{18}O$ is a little heavier (two more neutrons).

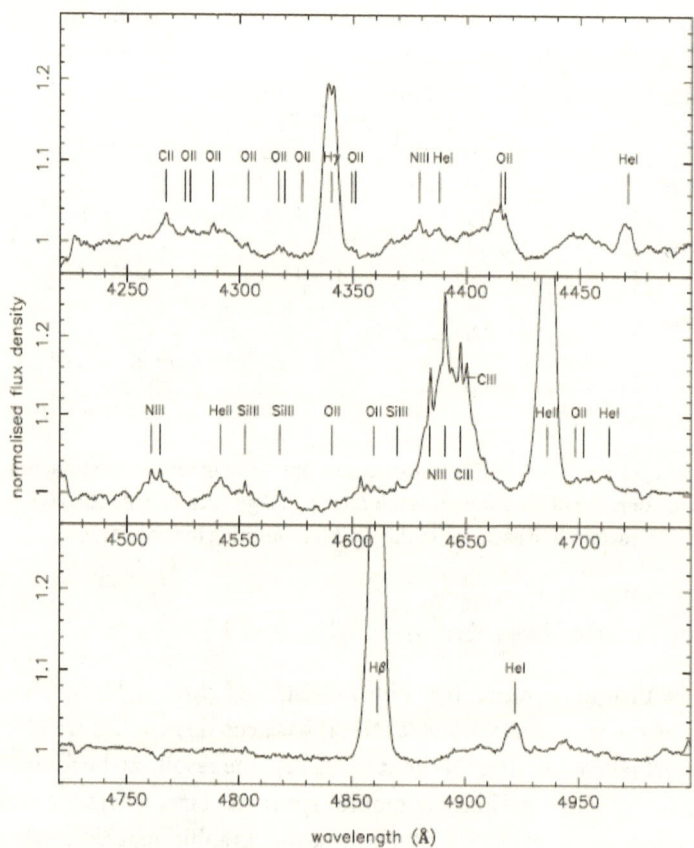

Figure 14: Portion of the spectrum [87] of Scorpius X-1, number 13 in table 2, page 49. A number of lines overlap precisely with the sun spectrum in figure 24. Wavelengths are indicated in angstroms (Å); 1 Å = 10^{-10} m.

Nine Thousand Light Years Away: Scorpius X-1

Consider, finally, Scorpius X-1, number 13 in table 2, more than nine thousand light years away.[6] Its spectrum has been analyzed at length, and part of it appears in figure 14 [87]. Look at figure 24 page 94, which displays the spectrum of our sun. Both spectra bear the very same line around 4,860 Å, produced by the hydrogen atom. There is hydrogen on Scorpius X-1. But there is also helium, oxygen, silicon, carbon, nitrogen—all betrayed by the

---

6. Scorpius X-1 has its own Wikipedia page.

lines we see in figure 14. The laws of nature are the same on Scorpius X-1 as they are on earth.

The Speed of Light Hasn't Changed Lately

I'm not going to go through every single object in table 2. For each one, you can find either a few or many reports on their atomic and/or molecular lines. These lines tell us that the world is the same "out there" as it is around us. More precisely, these lines tell us that the world *was* the same "out there" as it is around us *now*. If all these lines are found where we find them in the lab, then the laws of nature expressed by equation (2), or in Appendix A, were the same when light left the star. And if these equations were the same, then the speed of light was the same.

Just the three objects we've been through tell us the speed of light was the same nine thousand, twenty thousand, and thirty thousand years ago. In case you're wondering about the speed of light for the objects in the table *closer* than Scorpius X-1, you can easily find information on the last one, Cygnus X-1, a little more than six thousand light years away. Like Scorpius X-1, it has its Wikipedia page. Cygnus X-1 has been studied intensively since it was discovered in 1964, because it is an intense source of X-rays (the same ones that allow doctors to check your lungs or broken bones with radiography). I just counted 341 articles about it in peer-reviewed journals published up to May 2012. Hydrogen, helium, carbon, nitrogen, neon, calcium, oxygen, nickel, magnesium, silicon, iron—all have been spotted there [42, 68, 84], among many others. Known laws of nature are the same there also.

Spectroscopy, the science devoted to the study of spectral lines, is incredibly useful to astrophysicists. An analysis of the global sun spectrum gives its surface temperature—about 5,500 degrees Celsius. Spectral lines tell us about the velocity, with respect to us, of the place we're looking at, but also its chemical composition, its temperature, its gravity, and much more. If you're interested in knowing all the details, reference [44] is *the* book on that topic (warning: very technical).

What can be done for the sun has obviously been done for every single object we can observe up there. Stars, pulsars, star-forming regions, galaxies—literally millions of these have already been cataloged.[7] The vast majority are so far away that the parallax method does not work for them.

---

7. Just google "astronomical catalog." They are now on the Web.

The angles of figure 9 are so close to 90 degrees that our instruments can't measure the difference. If we can measure these angles for the objects closer than thirty thousand light years listed on page 49, this implies that the rest are *farther* away than that.

Beyond the twenty-six objects listed, we know of millions that are farther away. Reference [73] contains spectra from two stars farther than ten thousand light years away which are not in table 2. Such observations are now routine, and I had to stop registering them at one point. The table does not gather some rare occurrences of distant stars. It is just a glimpse over literally millions of objects that far, and farther, away. The atomic or molecular spectra displayed in the last pages are not some unique specimen hardly gathered by scientists. They are just a little sample of literally millions more.

So yes, the laws of nature have been the same over the last six thousand years, and so has the speed of light. Light we see from objects nine thousand, twenty thousand, or thirty thousand light years away did leave them nine thousand, twenty thousand, or thirty thousand years ago. As a direct consequence, our universe was already there nine thousand, twenty thousand, or thirty thousand years ago at the very least. This is an observed fact.

5

Radio Dating and Astrophysics

In a Nutshell

- An atom has a nucleus with electrons revolving around it. "Radio decay" is a natural process according to which some nuclei spontaneously change into others.
- *Assuming* that the rate at which this happens does not change with time, various nuclei act as "natural clocks." Some of them tell us that the world is much older than six thousand years. These rates are called "decay rates."
- The most famous of these natural clocks is carbon-14.
- What about the constant decay rates assumption? Decay rates obey the laws of nuclear physics. According to these laws, some decay rates can change when modifying the electron cloud surrounding the nucleus.
- Every change observed is in line with predictions. For the most dramatic one, experiments were even done to test the theory.
- But the laws themselves don't change with time. Still, we could have missed something. Would a given experiment confirming these laws *now* give the same result if it had been performed six thousand years ago?
- These laws predict, among many others, the decay rates of nickel-56 and cobalt-56. They also predict that titanium-44, aluminium-26, and iron-60 emit a very special kind of light when decaying. Both aluminium-26 and iron-60 are involved in dating techniques.

- Astrophysical events called "supernovae" allow monitoring nickel-56 and cobalt-56 decay rates. These supernovae occur much farther than six thousand light years. More than 2,600 of these have already been observed. Nearly twelve new ones are detected each month.
- The observed decay rates are the same as now. The very special light emitted when the other elements decay is equally detected exactly where expected from the laws we have now.
- The laws of nuclear physics were therefore the same more than six thousand years ago, when these events happened. And all the decay rates depend on these laws.
- Claiming that the laws of nuclear physics were the same in the past is not an assumption. Rather, it is an *observation*.

Radioactivity is the process according to which some atoms spontaneously change into others. Such atoms are called "unstable." The others, the "stable" ones, don't change into anything else. They just stay there. Because the rate at which radio decay occurs is constant (I will comment on this later), this phenomenon is frequently used as a dating technique. The carbon-14 method is certainly the most famous one, which is why I will explain how it works, in order to illustrate the radio dating method in general.

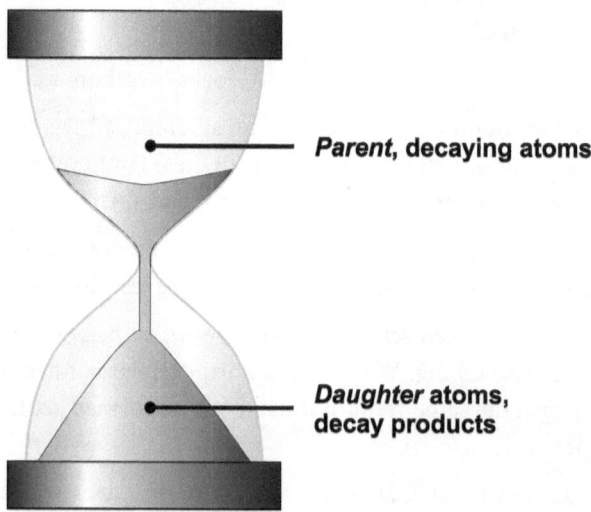

Figure 15: Radio decay can be compared to a sand clock. *Parent* atoms decay, becoming *daughter* atoms as time passes.

Radioactivity has been compared to a sand clock such as the one pictured in figure 15. The decaying atoms (the *parent* atoms) are represented by the sand on top, which progressively flows down to the lower chamber, where the sand now represents the decay product (the *daughter* atoms). By looking at the amount of sand on the top or in the bottom, you can tell how much time has passed since the clock was triggered, provided the following:

1. You know the rate—the famous "decay rate"—at which the sand flows from above.
2. This rate didn't change with time.
3. You know how much sand was there at the beginning.
4. No sand could escape or enter the clock in the process.

It turns out that some natural "sand clocks" indicate that they were triggered long ago—much longer than six thousand years. For reasons I will mention, point number two above is the one that needs further explanation. Here again, astrophysics allows us to check *now* some decay rates in the *past*. We saw earlier that gravitation has its laws. Then we talked about how the laws of atomic physics allow predicting atomic and molecular spectral lines. The decay rates we're now discussing rely on the laws of nuclear physics, namely, the physics of the atomic nucleus. If you're keen on knowing a little more about it, please see Appendix B.

The question of the constancy of decay rates is therefore related to the constancy of the laws of nuclear physics. Same nuclear laws, same decay rates—exactly like "same gravitational laws, same orbits" or "same atomic laws, same spectral lines." Here again, astrophysical observations allow us to confirm that these nuclear laws were indeed the same in the past.

But before we get there, I would like to explain how dating techniques work using the example of carbon-14. Then, I would like to clarify one important point about decay rate constancy: yes, it is possible to change it by modifying the electrons orbiting around the nucleus. This kind of change is completely understood by scientists. But a given atom will always decay at the same rate if you leave it alone. Simply put, decay rates can vary with *composition*, not with *time*.

Carbon-14 and Other Dating Techniques

Although I won't emphasize carbon-14 dating too much, I decided to cover it briefly because it is by far the most famous dating technique. There are various kinds of carbon atoms. Carbon-12 (12C), with six neutrons and six protons in its nucleus, is stable. There is also carbon-13, which has six protons and seven neutrons in its nucleus and is likewise stable. Finally, 14C has six protons and eight neutrons in its nucleus and is unstable, decaying to nitrogen-14 (14N).

Suppose you have a carbon bar initially made up of 14C and 12C in equal proportions, like the one on the left in figure 16 below. Wait about 5,730 years, and half of the 14C will have turned to 14N (as Appendix B makes clear, this is a β⁻ decay). Wait another 5,730 years, and half of the remaining 14C will have turned into 14N. This process repeats itself. This phenomenon has been extensively observed in laboratory [33]. The period of 5,730 years is called the "half-life" of 14C and has been measured with an accuracy of thirty years (0.5 percent accuracy).

From the details of the decay process described above, the amount of 14C remaining after x years can be determined quite precisely. The formula is as follows:

$$N_{14C}(x) = N_{14C}(0)\exp\left(-\frac{x/\ln 2}{5,730}\right) \quad (3)$$

Here x is the number of years, $N_{14C}(0)$ is the initial amount of 14C in your sample, ln 2 ($\approx$ 0.693) is the natural logarithm of 2, and "exp" is the exponential function. If the formula above is Greek to you, don't worry—you won't miss my point. The important point is that it exists.

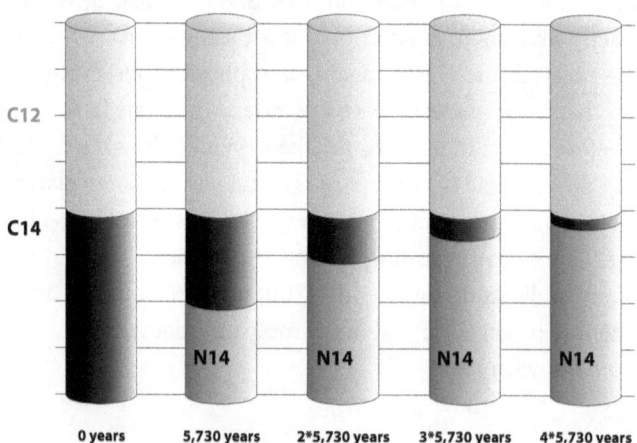

Figure 16: Carbon-14 decay to nitrogen-14. Every 5,730 years, half of the remaining carbon-14 decays to nitrogen-14.

We finally come to the core concept of 14C dating. We human beings are partly made of carbon. So are trees, animals, plants, and every living organism on earth. Where does this carbon come from? In the case of plants, it comes from the fixing of atmospheric carbon through photosynthesis (a kind of plant respiration). When herbivores eat plants, they indirectly use atmospheric carbon to build their bodies. And when we human beings eat plants or animals, we also recycle atmospheric carbon. Now, carbon in the atmosphere is for the most part 12C, with a tiny fraction of 14C. For chemical reasons I won't emphasize, organisms assimilate 14C in exactly the same way as 12C. As a consequence, the ratio of 12C to 14C in the atmosphere is the same as in my body, in the tree I see through my window, or in my neighbor's dog. As long as these entities are connected to the atmosphere, as long as they eat or breathe, this ratio remains the atmospheric one—until the connection stops, that is, until we die.

When I die, my stock of carbon will no longer be renewed by atmospheric carbon. While the 12C in my bones will remain 12C (it is stable), the few atoms of 14C there will decay to 14N. Note that while I was alive, the 14C in my body was already decaying to 14N. But my connection with the atmosphere would continually bring in fresh 14C, which kept my ratio of 12C to 14C constant. Now that the link is broken (because I'm dead), my 14C is no longer being replaced and begins vanishing according to the process depicted in figure 16.

Knowing my 12C/14C body ratio the day I die, I, or preferably someone else, can later relate the time of my death to this very ratio. Let's look at an example: the current 12C/14C ratio is one part per trillion. In other words, out of 1,000 billion carbon atoms in the atmosphere, only one is 14C. Fortunately, there are so many atoms, even in a single drop of water, that 14C atoms are numerous despite so small a ratio. Suppose I die today (well, right after I finish this book). My bones contain 14C in atmospheric proportions, which start decaying to 14N without being replaced. Suppose they bury me and that, at some distant point in the future, an archeologist finds my skeleton. He analyzes the 12C/14C ratio in my bones and finds one 14C out of 2,000 billion 12C. If he knows that this ratio is half what the ratio was on the day I died, he can readily deduce that I died some 5,730 years before. If he finds only one 14C out of 4,000 billion 12C, he'll compute that I died 11,460 years ago (2 · 5,730). The elapsed time since an organism died can therefore be deduced from the 12C/14C ratio, provided the initial ratio is known. Note that because modern paper as well as old papyri or parchments are made of plants, the age of the plants they came from can be estimated in the same way.

The atmospheric carbon ratio is therefore a key issue in dating. To start with, you may have wondered why the amount of 14C in the atmosphere does not fall to zero with time, as it will in my bones after I die. The reason is that 14C atoms are continually produced in the atmosphere by particles coming from space. Thus the 14C production rate may be affected by extraterrestrial events. Measuring the ratio today and assuming that it has always been the same in the past yields an initial dating method that will require corrections.

It is nevertheless very interesting to compare the date attributed to a number of historical artifacts by the 14C method *without* such corrections. This was first done in an article titled "Age Determinations by Radiocarbon Content: Checks with Samples of Known Age" [3]. Table 3 shows some of the results. The method followed by these authors was very simple: collect historical artifacts of historically known age and compare their age with the outcome of the 14C dating method assuming a constant atmospheric 12C/14C ratio.

Sample	Known age	Carbon dating
Tree ring	1,372 ± 50	1,000 ± 150
Ptolemy	2,149 ± 150	2,300 ± 450
Redwood tree	2,928 ± 52	3,005 ± 165
Zoser Pharaoh	4,650 ± 75	4,750 ± 250

Table 3: Carbon dating of samples of historically known age [3].
Ages are given in years before 1949.

Two important comments can be made regarding table 3. First, the carbon dating method has a margin of error, but this margin is less than 20 percent. Second, the 14C date is in agreement with the historical one. As already mentioned on page 16, the Dead Sea Scrolls, a collection of Old Testament fragments discovered between 1946 and 1956 near the Dead Sea, in Israel, have been dated using this technique to a few centuries BC. They pushed back the date of the oldest known Old Testament original manuscript by more than one millennium!

Comparison with history is one way to calibrate 14C dating: you compute the age of an artifact assuming that the 12C/14C ratio was the same as it is today, then you compare with historic dating, and you update the initial ratio parameter accordingly. But how do you date an artifact before "history"? A widely used method is that of counting tree rings: you find some very old trees and just count the rings. Then you date these rings by the 14C technique. Since one ring has been added every year, you can compare the calculated 14C age with the real one.

This method and others allowed for the calibration of this dating technique up to 20,000 years BC [88, 89, 78]. Beyond this point, the 14C technique becomes useless because as you divide the quantity of 14C by two every 5,730 years, you end up with nothing detectable after more years.

So 14C can tell you about a world older than six thousand years. But 14C is completely useless beyond a few tens of thousands years. How do you use the radioactivity technique to find objects older than that? Simply by using atoms decaying *slower* than 14C. Aluminium-26, for example, has a half life of 0.7 million years. Uranium-235 has a half-life of 0.7 *billion* years. Reference [1], for example, explains how lead dating reveals that some meteorites are 4,564 million years old. Dating methods using these atoms work in the same way the 14C method works. The only difference is that you exchange thousands of years for millions, or even billions. Instead of dividing your initial stock by two every 5,730 years, you do so every 0.7 million years (for aluminium-26) or 0.7 billion years (for uranium-235). You can therefore wait much longer before there's nothing left to detect.

Parent	Daughter	Half-life	Type of decay
Aluminium-26	Magnesium-26	0.7 million	β^+
Iron-60	Nickel-60	2.6 million	β^-
Rubidium-87	Strontium-87	4.9 billion	β^-
Samarium-147	Neodymium-143	10.6 billion	α
Lutetium-176	Hafnium-176	37.8 billion	β^-
Rhenium-187	Osmium-187	42.3 billion	β^-
Platinium-190	Osmium-186	450 billion	α
Thorium-232	Lead-208, helium-4	14 billion	α
Uranium-235	Lead-207, helium-4	0.7 billion	α
Uranium-238	Lead-206, helium-4	4.4 billion	α

Table 4: Some useful parent/daughter dating couples. The decay chain for iron-60 is as follows: iron-60 → cobalt-60 → nickel-60. Reference [45] (table 1, page 2) lists sixteen more parents. See Appendix B for the "type of decay" column.

Table 4 lists ten useful parent/daughter couples. Reference [45] presents a list of twenty-six. Indeed, we have more than one single technique. In the next sections I will focus more on the stability of the decay rates than on which atoms are used in terms of the problem to address. If you want to know more about exactly how radio dating techniques are used, references [45] and [102] will help, as will the Wikipedia page on radiometric dating.

Decay Rates Can Vary

Let me now go back to the four assumptions given on page 63 and focus on one of them. As a reminder, these four assumptions underlying any radio dating technique are as follows:

1. You know the rate at which the parent atom decays.
2. This rate didn't change with time.
3. You know how many parent atoms were there in the beginning.
4. You have a closed system—parents and daughters can neither enter nor escape.

Assumption 1 does not pose any problem. You just have to do the experiment in your lab and measure the rate at which the parent turns into its daughter. Yet, you may wonder: how can a half-life of one billion years be *measured*? Of course, you can't monitor your atoms for a billion years to determine when half of them will be gone. Fortunately, the number of atoms you have even in a minuscule sample of matter is enormous. For example, a single drop of water contains about 10^{21} (one followed by twenty-one zeros) oxygen atoms. How big is this number? Suppose you take a drop of water and put a mark on each oxygen atom it contains. You then release it in the sea and wait a few thousand years for your drop to mix with *all* the water available on the planet. By this time, you go wherever you can find water on earth and take a one cubic meter sample of it. It should hold about five of the oxygen atoms you marked! That's how big 10^{21} is. One drop is enough to "pollute" all the water on earth.

So when you're sitting in your lab, doing an experiment to measure a billion years half-life, you don't have to wait a billion years. If half of the atoms are to be gone by, say, one billion years, it takes a much shorter time for one atom out of one million (instead of one out of two) to vanish. And you have so many atoms that the departure of one out of one million is definitely detectable. You therefore measure the time it takes for a tiny little fraction to decay, and you deduce the time it would take for half of them to do the same. Obviously, you here need point 2 of the list above, since you're observing your atoms for a few hours instead of a billion years. This is why the rest of this section is devoted to this very point.[1]

What about assumptions 3 and 4? Although less obvious, they can be tackled. Regarding assumption 3, for example, a radio dating technique such as isochron dating allows us to get rid of the initial amount problem by tracking various ratios instead of one. Assumption 4 can likewise be taken care of when dealing with samples trapped in rocks. Reference [102] gives a thorough review of these issues.

Assumption number 2 is therefore the key, not only to present half-life measurements but also to the very reliability of the method. If the rate at which sand falls from above changes *with time*, how can I know how long it's been inverted?

As is the case with the speed of light, the half-life stability issue was raised long ago by physicists. The Italian physicist Emilio Segrè, who won the Nobel Prize in physics in 1959, first wrote about this possibility in 1949 [85], while a French team did the same in the very same year [8]. Is it possible to alter the rates we've talked about? Interestingly, the answer is *yes*, at least for some elements. But it is *not* easily achieved. Let me just quote part of the first paragraph of reference [32], which reviewed the knowledge in this field in 1972:

> Early workers tried to change the decay constants of various members of the natural radioactive series by varying the temperature between 24 K and 1280 K, by applying pressure of up to 2,000 atm,[2] by taking sources down into mines and up to the Jungfraujoch,[3] by applying magnetic fields of up to 83,000 Gauss,[4] by whirling sources in centrifuges, and by many other ingenious techniques. Occasional positive results were usually understood, in time, as the result of changes in the counting geometry, or of the loss of volatile members of the natural decay chains. This work was reviewed by Meyer and Schweidler, Kohlrausch, and Bothe. Especially interesting for its precision is the experiment of Curie and Kamerlingh Onnes, who reported that lowering the temperature

---

1. A look at www.nndc.bnl.gov/chart will reveal that rates have now been determined with great precision for almost every known unstable nucleus.
2. That is, two thousand times the atmospheric pressure.
3. A col 3,471 meters high in Switzerland.
4. This is 160,000 times stronger than the earth's magnetic field moving your compass.

of a radium preparation to the boiling point of liquid hydrogen changed its activity, and thus its decay constant, by less than about 0.05%. Especially dramatic was an experiment of Rutherford and Petavel, who put a sample of radium emanation inside a steel-encased cordite bomb. Even though temperatures of 2,500 ºC and pressures of 1,000 atm were estimated to have occurred during the explosion, no discontinuity in the activity of the sample was observed.

It is clear that in line with what we've seen in chapter 3 page 35, scientists have been trying *hard* for more than sixty years to break the constancy of the decay rate. And they've been successful at it. In 1972, when the article excerpted above was written, the largest departure from the normal decay rate[5] produced in an experiment was 3.5 percent for niobium-90. Reference [25] tells how this was achieved by carefully changing the chemical environment of the element.

But before we draw conclusions about the decay rate variability, let me mention a spectacular effect observed in 1996. We just saw how people have been trying to produce variations of the decay rate under extreme conditions. Heating, cooling, pressure, magnetic field were tried, and extreme cooling was found to produce a 0.05 percent change in the radium decay rate, as quoted above. Later, a clever design of chemical environment for niobium-90 produced a 3.5 percent decay-rate shift [25]. Why was chemistry more successful than temperature, pressure, and so on? Because these temperature or pressure conditions were not high enough to affect the orbits of the electrons around the nucleus.

The atom is a little world in itself (see figure 22, page 91). As long as the electrons keep moving in the same way, this little world remains strictly unchanged and will continue working the same. The closer you come to the atom, the more likely you are to change something. And the first step in initiating a change consists in "touching" the electrons. That is precisely what chemistry is all about—putting atoms in contact and having their electrons, not their nuclei, interact.

Now, what if you take away *all* of the electrons from an atom? That's a tremendous change. This is exactly what some scientists did in 1996 [12]. The rhenium-187 element (75 protons, 112 neutrons) normally decays[6] to the stable osmium-187 (76 protons, 111 neutrons). As long as it

---

5. β^+ decay. See Appendix B.
6. β^- decay. See Appendix B.

is neutral—that is, surrounded by its 75 electrons—its "neutral" half-life is about 42 billion years. In a very delicate experiment, researchers removed every single one of these 75 electrons and measured the half-life of the naked atoms. Result: about 33 years! Undressing the atom divided its half-life by more than one billion! Was this a surprise? No, because it was in line with what could be expected based on the laws of nuclear physics. Indeed, this experiment had been performed to *check the theory*.

So, is every single observation perfectly understood? Reference [38], published in 2009, summarizes the possible detection of *unexpected* decay-rate variations. Let me here quote a sentence from the same authors, in reference [39], titled "Evidence for Time-Varying Nuclear Decay Dates: Experimental Results and their Implications for New Physics":

> Notwithstanding the impressive body of evidence supporting the conventional view that the decay rate $\lambda = \ln 2/T_{1/2}$ of an unstable isotope is an intrinsic property of that isotope, there are growing indications of small [$O(10^{-3})$] time-dependent variations in the decay rates of some nuclei.

What is the meaning of this [$O(10^{-3})$] symbol? It means the shifts possibly observed in the decay rate of *some* elements amount to a few tenths of a percent. According to this article, there could be some correlations between the decay rate of some elements and the sun-earth distance or the sun's activity. Interestingly, the issue is far from being settled, as some articles have been published since 2009 definitely *refuting* the correlation [69, 49, 7]. There is therefore an ongoing debate in the scientific community, and many, including me, will stay tuned to see how it evolves. Here again, as was the case for possible small variations of the "fine structure constant" discussed earlier on pages 54–55, two remarks can be made:

1. The observations are much debated. For example, the most recent work on the fine structure constant variation [99] concluded that "within the uncertainties, [our observations are] consistent with no variations of the fundamental constants."

2. The few *hypothetical* unexplained observations are far from being able to bring the age of the universe back to six thousand years. For half of your uranium-238 to decay in a few thousand years, you somehow need to reduce its half-life from 4.4 billion years to just a few thousand (divide by one million). This is not a *small* alteration of the laws of nuclear physics; this is a *huge* one. A miracle.

Because of Laws We Know... and Not with Time

Yes, decay rates can change—if you change the atom by playing with its electrons. And we know *why*. But the same atom will always decay at the same rate. The only unexpected observations of decay rate variations are 1) fiercely debated and 2) of the order of 0.1 percent anyway.

Radioactive atoms would be unreliable guides for dating if half-life changes were not understood, if they were random, unpredictable in time and kind. But this is absolutely *not* the case. Would you claim your car cannot carry your family safely because its speed can change? No, precisely because you control that speed.

Decay rate changes are not unpredictable, nor are they something that the known laws of nature completely miss. I won't write here the equations governing physics at the nuclear level because they are far too long (besides, I have already beset you with too many equations). But there are equations from which you can calculate decay rates and everything else related to nuclear physics. These equations are harder to solve than the ones we saw previously, which is why the agreement with experiments is less spectacular. But nuclear power plants and nuclear medicine work, and not only by chance. I will now review the decay rate variations noted previously and show why they are no surprise.

- Why did Emilio Segrè look at changes in the decay rate of beryllium-7 in 1949? Because beryllium-7 undergoes a form of decay called "electron capture," in which the nucleus absorbs one of the electrons orbiting around it. He simply reasoned that if this decay depends on the electrons, then chemistry could change it. The relative shift has been measured since then and found to be around 1 percent at best [74].

- Radioactive elements decay according to several "modes." For one of them, the so-called α mode, decay rates don't change. And we know why (as explained in Appendix B).

- The dramatic half-life reduction by one billion times of the fully stripped rhenium-187 is *understood*. Although spectacular, the result was expected and *quantified*. The expected value was fourteen years, versus the thirty-three years measured. Not too bad, when you think about the starting point. Is our nuclear physics wrong because it predicted fourteen years although the actual number is thirty-three? No.

The discrepancy has two origins here. First, these half-life measurements are very difficult, and error bars important (± 2 years). Second, the theory itself is mathematically very involved, and it is difficult to compute precisely what it predicts.

This huge half-life reduction was a theoretical *prediction* made in 1987 [90], nine years before the 1996 experiment. This explains how the authors of the experiment decided upon this atom. Theory had told them where to look in the first place.

- If decay rates can change when playing with the electronic cloud, how can we know that the atoms we use for dating were stable in the past? Which conditions would be required for rhenium-187 to lose all of its electrons, for example? Take a bunch of rhenium-187 atoms and put them somewhere. How hot should this "somewhere" be to fully strip rhenium-187 atoms of their 75 electrons? At the very least, two billion degrees.[7] There are very few places like this in the universe. The center of the sun is "only" fifteen million degrees.

  Besides this extreme example, it is never easy to remove an electron from an atom. The atom wants to keep it. The *easiest* atom to deprive of its electron is hydrogen. Still, it takes a "somewhere" about 10,000 degrees hot to kick electrons out of some hydrogen atoms. By comparison, the hottest lavas in volcanoes are cooler than 1,500 degrees.

- Of course, people have been trying to see what happens when undressing (electronically) atoms other than rhenium-187. Reference [57] reports experiments on neon-19, iron-52 and iron-53, manganese-52, praseodymium-140, promethium-142, and tantalum-168 (all β⁺; see Appendix B). The measured half-lives differ significantly from the dressed, neutral atom in all but iron-53 and manganese-52. But the shift never reaches the rhenium extremes. The largest relative

---

7. For those interested, here's the calculation: The energy needed to take out the last electron would be $13.6 \, Z^2$ eV, with $Z = 75$. Assuming this last electron perfectly screens the nucleus, the energy needed to take out the penultimate electron is $13.6 \, Z^2$ eV, with now $Z = 74$. Forgetting about what is needed to remove the first 73 electrons (!), we find the total ionization energy is at the *very* least $E = 13.6 \, (75^2 + 74^2) = 151{,}000$ eV. For the gas to be hot enough to completely strip the rhenium-187 atoms, its temperature T must satisfy $k_B T > E$, where k_B is the Boltzmann constant. Putting the numbers gives $T > 1.75 \cdot 10^9$ Kelvin, which I round to two billion degrees.

half-life shift is found for praseodymium-140, which goes from 3.4 (neutral) to 7.3 minutes (naked). The naked half-life—that is, without electrons—is always theoretically computed with a precision better than 20 percent except for manganese-52, where it reaches almost 50 percent. A similar experiment was conducted recently (see reference [5]) for iodine-122, with 53 electrons to remove. No significant half-life shift was found, as *expected* from theory.

Besides rhenium-187, the elements mentioned have a really short half-life. Iron-52 has the longest one, of only 8.28 hours. They are therefore completely irrelevant for dating, as they vanish after a few days.

We know about 3,000 stable and unstable nuclei. If we double this number (taking into account both the neutral atom and the naked atom), we get nearly 6,000 atoms. Clearly, theory has not been tested against every single one of them yet. But the known laws of nature describe correctly the outcome of every experiment that has been performed so far. These laws are the ones we saw in the previous chapter, together with the ones that rule the nuclear world and that are at work in nuclear medicine, nuclear power plants, or nuclear weapons.

The atoms listed in table 4 have obviously been tested and belong to the realm of what science explains well. The reliability of dating methods eventually boils down to the issue discussed in chapter 4, from page 53: have the laws of nature we know always been the same, or not?

The previous section on objects farther than six thousand light years away partly answered that question. From the light spectrum we observe from millions of objects farther than this, we know the laws of quantum mechanics and electromagnetism (among which is the speed of light) were the same up there, when the light now reaching us was emitted. Equations (2) were valid on these stars, pulsars, nebulae, or whatever, when light left them more than six thousand years ago. There are corresponding laws of *nuclear* physics, which are too long to include here. But still, the question arises: what evidence do we have that these laws were the same in the past? Here again, important clues can be gathered from astrophysics.

Nuclear Physics and Astrophysics

Supernovae

In 1572, the Danish astronomer Tycho Brahe noticed the sudden appearance of a very bright point in the sky. He later wrote an essay on it, titled *Concerning the New Star, never before seen in the life or memory of anyone.* He was precise enough so that we know where to look to observe the object he found. Tycho's supernova, or SN 1572 in modern astronomical lingo, is what we now call a Type Ia supernova. Supernovae are exploding stars, and Type Ia supernovae are just a subclass of them. The remnant of Tycho's supernova can still be observed today in the form of a bright disk in the sky. Its distance to us is calculated at between 8,000 and 9,800 light years. A list of supernova is available online;[8] according to my count, the list contains 2,673 Type Ia supernovae. With current observation techniques, 290 supernovae were detected in 2011, of which 151 were Type Ia. As of May 31, 2012, 93 supernovae had been detected in 2012, with 46 Type Ia among them. We are averaging 2.6 Type Ia supernovae per week. These are anything but rare events.

Why are they important? First, because they are very far from us. Since they remain visible for a long time (Tycho's SN 1572 has been observable for 440 years), there is plenty of time to get a parallax. In general, angles 1 and 2 (as depicted in figure 9 page 47) are too close to 90 degrees to calculate any distance. This implies that these stellar explosions occur at distances greater than the objects listed in table 2, page 49. For example, the distance to the quite famous supernova SN 1987 (similar to Type Ia) has been measured by various methods I won't describe, and found to be approximately 160,000 light years away [48, 70].

Second, once the supernova has appeared like a new star in the sky, its light can be observed in terms of its wavelength, but also in terms of time. You can look at it today and observe the spectral lines. Then you can look again next week and see what happened to these lines. If you plot in terms of time the intensity of light received in a given frequency range, you obtain something like figure 17 (adapted from reference [37]). Here, light curves for five events have been recorded, and all five are remarkably similar. All Type Ia supernovae light curves look like this one—same initial steep slope

---

8. See www.cbat.eps.harvard.edu/lists/Supernovae.html.

from the maximum, then same break some twenty days from the top, and slower final vanishing on the same timescale.

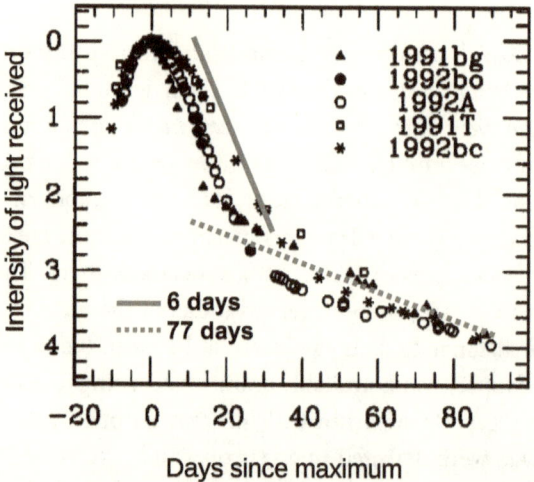

Figure 17: Intensity of light received as a function of time for five different Type Ia supernovae observed in 1991 and 1992. I added the slopes expected from half-lives of 6 days and 77 days. (Figure adapted from reference [37].)

Obvious question: what is happening? Here, theoreticians gave the clue to observers. Our sun provides energy resulting from nuclear fusion reactions occurring in its center.[9] These reactions produce an enormous amount of heat, which builds up a pressure opposing gravity. Gravity wants to shrink the sun, but internal pressure prevents it, and the sun's radius results from a balance between these two. Now, these reactions need "fuel" in the same way that fire needs firewood. When you've burned all the firewood, well, the fire goes out. So what will happen when all the nuclear fuel is spent? Heat generation will stop and will no longer be able to oppose gravity. The sun will contract, increasing the core pressure. This compression will reheat the core and trigger nuclear reactions that were impossible before. Because these reactions are fusion reactions, larger and larger nuclei are produced. Right now, the sun is burning hydrogen to form helium. We go from a one-proton nucleus (1p) to a two-proton nucleus (2p). Then the

---

9. A fusion reaction *merges* two nuclei into a single one. For example, two hydrogen atoms with one proton and one neutron each, can merge to form an helium nucleus, with two protons and two neutrons. Fusion is the opposite of fission, which powers nuclear power plants.

sun will burn helium (2p) to form carbon (6p), and then oxygen (8p)—until you get to heavy nuclei such as nickel (28p). The story is in fact more complicated than that. But the main points are there.

An important feature is that if you start from hydrogen (1p, 1n) or helium (2p, 2n) and grow your nuclei, feeding them with hydrogen or helium, you can only generate elements with an even number of protons and neutrons. For a reason you may understand after looking at figure 25 of Appendix B, these atoms are unstable.[10] As a consequence, your star should generate huge amounts of radioactive matter (don't tell Greenpeace), and when it explodes, you should see it.

Following this kind of scenario, and putting together what they knew about gravity, electromagnetism, nuclear physics, and so on, theoreticians predicted that some stars should generate huge amounts of nickel-56 (28p, 28n). As expected, nickel-56 is radioactive. It decays to cobalt-56, which in turn decays to the stable iron-56. The predicted decay chain, with the expected half-lives, is as follows:

Nickel-56 → cobalt-56 → iron-56
6 days 77 days

When the star explodes, all its radioactive nickel-56 is released. Radio decay releases a lot of energy. The energy heats the environment,[11] and this heat can be detected because it produces light. Because nickel-56 has a half-life of six days, the light produced by its decay should also decay at such a rate. This is exactly what is observed in figure 17, where the plain line shows the predicted decay slope corresponding to a six-day half-life.

Quickly, then, all the nickel-56 turns to cobalt-56, which starts decaying. This time, the light curve should evolve with time according to such a decay law. Again in figure 17 we see this very trend, with the dotted line displaying the theoretical decay slope corresponding to a 77-day half-life.

---

10. You can even find they will be β^+ unstable because they will all lie on the straight black line in the figure.

11. Radioactivity produces heat. This is why the Fukushima nuclear reactors had to be cooled after the March 2011 tsunami destroyed the cooling system.

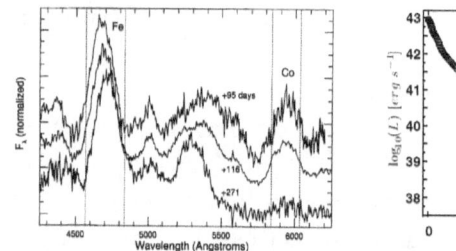 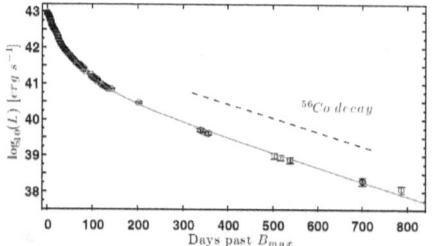

Figure 18: *Left*: Iron (Fe) and cobalt (Co) lines days after the luminosity peak of supernova SN 1981B. (Adapted from reference [54].) The same reference reports similar results for twelve additional supernovae. *Right*: Long-term luminosity of SN 2003hv together with a straight line indicating the slope expected from Co-56 decay. (Adapted from reference [56].)

With this clue about what to look for, observers went after cobalt and iron lines. Figure 18 (left) shows the lines corresponding to cobalt and iron in the days after the luminosity peak of supernova SN 1981B. If you compare the relative amplitudes of the iron (Fe) and cobalt (Co) lines, you'll see how the ratio cobalt/iron gets lower with time, on the timescale expected from the 77-days half-life. The same reference reports similar results for twelve additional supernovae. Figure 18 (right) shows the long-term luminosity of SN 2003hv, another supernova, together with a straight line indicating the slope expected from cobalt-56 decay. Direct observation of cobalt-56 decay has also been reported for supernova 1987A in reference [95].

As it decays, cobalt-56 emits light with very special properties. It is a kind of light called γ ray, with very high and well-determined frequencies expected at 846.77 and 1,238.288 keV. These "keV" are just units of energy, and the amount of keV indicated is the one carried by each particle of the emitted light.[12] By observing supernova SN 1987A in the proper frequency range, researchers detected γ rays at 843 and 1,238 keV, evidencing again the presence of cobalt-56 decay [61].

There are therefore observational evidences for the nickel-56 → cobalt-56 → iron-56 decay chain in Type Ia supernovae. We observe the atomic lines, we observe the light emitted during the decay, and we can measure the half-lives involved. This decay chain was working in the same way tens of thousands years ago as today. And if it was, then the laws of nuclear physics were the same at that time.

---

12. Technically, we'd say cobalt-56 decays producing γ photons at 846.77 and 1,238.288 keV.

Titanium-44

Titanium-44 has twenty-two protons and neutrons. It is unstable and would like to get a larger neutron/proton ratio. So it decays[13] to calcium-44, with twenty protons and twenty-four neutrons. Doing so, it emits light at 1,160 keV. This very light has been detected from a supernova remnant called Cassiopeia A, the result of a star that blew up perhaps at the end of the seventeenth century. This object has been measured at a distance of 3,400 parcecs [75], that is, a little more than eleven thousand light years. In recent years, a number of telescope-satellites have been launched to observe the sky around the frequency where this kind of light can be found (the γ-ray band). One of them, the Compton Gamma Ray Observatory, looked at Cassiopeia A in search of this kind of light. Reference [51] reports the detection of the 1,160 keV photons, evidencing titanium-44 decay.

Reference [83] reports that this same light from the decay of titanium-44 has also been spotted in the supernova remnant RXJ0852.0-4622, less than 3,200 light years away from us [4].

From the evidence presented in the first chapter, we know that the speed of light was the same 3,200 and 11,000 years ago. Not only was there titanium-44 at that time, but it was emitting exactly the same light that it emits today when it decays.

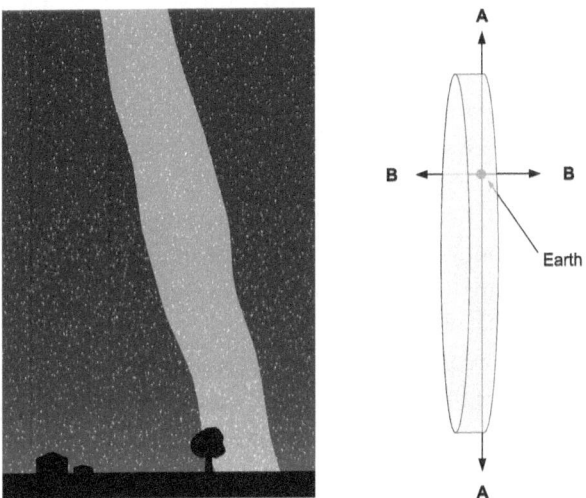

Figure 19: *Left*: Night sky featuring the Milky Way, our galaxy.
Right: Sketch of the Milky Way, and of the position of the earth within it.

---

13. β^+ decay.

Aluminium-26

Aluminium-26, with thirteen protons and neutrons, is listed in table 4. It decays to magnesium-26, which has a half-life of 0.7 million years. While doing so, it emits a photon with a very well-defined energy of 1,808.6 keV.

Allow me a quick digression in order to set the stage for my point. First, read Genesis 15:5:

> [God] took [Abraham] outside and said, "Look up at the sky and count the stars—if indeed you can count them . . ."

Then do what Abraham was instructed to do: look up at the sky. Depending on where you are, and on the night you choose, you could see something as magnificent as the picture in figure 19 (left). Today, city lights make it difficult to enjoy this beauty. But the spectacle must have been familiar to Abraham. Of course, there are stars in every direction you look. But in some directions you see a larger number. A band roughly vertical on this sketch apparently contains many more stars. With a telescope, you could see that, indeed, this band is made of millions of stars. The horizon prevents us from seeing more of it. By repeating the observation at various times of the year, you could see that this region crowded with stars envelops the earth.

Simple observation with the naked eye tells us that we are somewhere inside a disk-like shape full of stars, such as the one depicted on figure 19 (right). Look in direction A and you see far more stars than in direction B. This disk is called the Milky Way—the galaxy we live in. Greek mythology proposes that it was formed when baby Heracles, son of Zeus, spit out milk from Hera's breast.

Figure 19 (left) depicts the Milky Way in visible light. What if you took a similar picture, but looked only at the light aluminium-26 emits when it decays or other kinds of appropriate light? The results are pictured in figure 20, which requires an explanation.

In the top left corner, you find the profile of the aluminium-26 line toward the center of the Milky Way twenty thousand light years away [30]. The sketch in the top right corner shows how to interpret the "galactic longitude" scale involved in the other graphs. It is simply the direction in which you look. The picture labeled "Aluminium-26" is the intensity of that line for many galactic longitudes [53]. For each longitude, many latitudes have been explored. Longitude is then on the horizontal axis. But on the vertical axis, instead of the latitude, is plotted the slight frequency shift observed with respect to 1,808.6 keV. Why? Because this shift tells us about the velocity of the emitting region with respect to us, in exactly the same way radar guns catch the speed of your car on a highway, as explained earlier. Note that the shift is really small, as it ranges from -1 to 1 keV for our line at 1,808 keV.

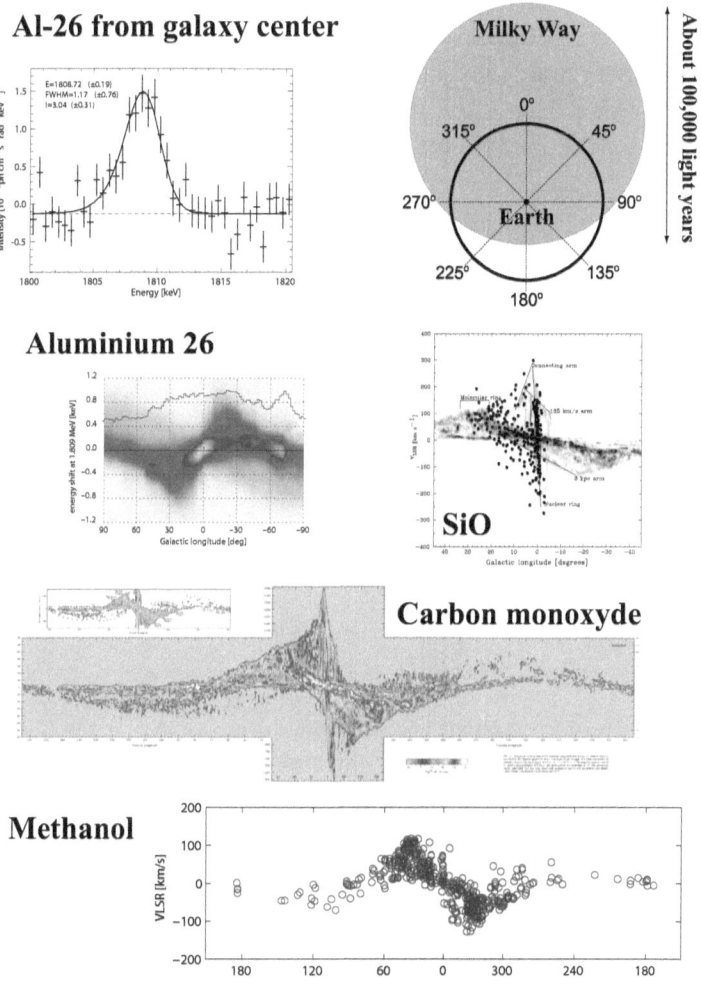

Figure 20: *Top left*: Aluminium-26 line from the galaxy center [30], and its galactic profile [53]. Other pictures: see text.

What are the other pictures? Obviously, they resemble the Aluminium-26 one. They all are pictures of our galaxy taken selecting very special *molecular* lines. We find here carbon monoxide [27], methanol [71], and silicon monoxide (SiO) [65].

These four pictures are remarkably coherent in the story they tell us: not only do they show that the physical world works the same throughout our galaxy, but they also show that our galaxy is *spinning*! Look toward the right of the galactic center, and everything is flying away from you. Look now toward the left, and everything is flying toward you. Note that velocities are higher toward the center than on the far left and right. The galactic disk is spinning. Such is the coherent message we get when looking at it, whether we ask atomic physics or nuclear physics.

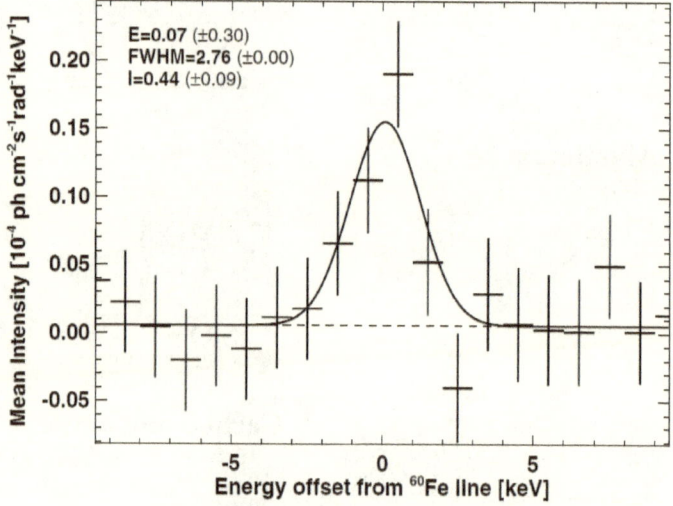

Figure 21: Superimposed iron-60 lines at 1,173 and 1,333 keV from the galaxy center [97].

Iron-60

Iron-60, also present in table 4, emits light at 1,173 and 1,333 keV during its decay. In reference [97], scientists measured the lines pictured in figure 21 from the center of our galaxy, approximately twenty thousand light years away.

These iron lines observed in the laboratory on earth are equally detected looking at the center of the galaxy. There is iron-60 near the center of our galaxy. And it decays in the same way it does on earth.

The Laws of Nuclear Physics Haven't Changed Lately

Do decay rates change with time? Do the laws of nuclear physics change with time? As was the case with the laws of atomic physics, the answer is *no*.

Several thousand years ago, nickel-56 and cobalt-56 were decaying at the same rate as today. We have a huge number of observations. Far more than six thousand years ago, cobalt-56, titanium-44, aluminium-26, and iron-60 were decaying, emitting the very same light that they emit today. Note that both aluminium-26 and iron-60 are listed in table 4 as atoms frequently used in dating. A number of ground-based or space-based instruments capable of detecting the light emitted during nuclear reactions are now operational. "Nuclear astrophysics" is a quickly developing field of physics,[14] and data are pouring in.

If the laws of nuclear physics had changed in the past, these decay rates, nuclear lines, and so on would have changed as well. Remember, a distinctive feature of the laws of physics is precisely that just a few equations govern every phenomenon. Atomic physics and astrophysics taught us that the speed of light was the same tens of thousands years ago, because the laws of atomic physics were the same. Here, astrophysics tells us that the laws of *nuclear* physics also were the same tens of thousands years ago. So were the decay rates.

Radio-dating techniques "assume" that decay rates did not vary in the past. How robust is this assumption? Indeed, it is extremely robust, because it is *not* an assumption but an observation. So is the universe more than six thousand years old? Observations and natural laws say "yes, definitely." As stated in the introduction, there is no *natural* way to reconcile current astrophysical observations with a young, six-thousand-year-old universe. You could assume that God intervened at one point, miraculously speeding up light and decay rates, or performing whatever other miracles you can think of. Of course he could do! But claiming that science is bordering the limit of our knowledge in this matter, or that "we don't know for sure," or even that there are scientific "evidences" for a young universe, is utterly wrong.

---

14. It even has its dedicated journal, the *Virtual Journal of Nuclear Astrophysics*, which gathers peer-reviewed papers on this topic.

Conclusion

Many Christians, including me, tend to ignore God's hand in everyday events. We like miracles. We like the unexpected. It is extremely tempting to think God has nothing to do with what science understands. This may be the only point both believers and atheists agree on. But Jesus does *not*. God causes his sun to rise. God feeds the birds. God sends the rain . . . I wish I could be spiritual enough to really *see* God's hand in these events, as I would for a miracle. Jesus was, and he told us. Everything, unexpected or not, understood or not, miracle or routine, is caused by God.

So why fear scientific explanations? We should be able to look freely at the teachings of science on any topic without feeling that our faith is being threatened. Do you think the sun rises because the earth rotates? I guess you do. Yet, Jesus says *God* causes it. Feel free therefore to simply enjoy what science has to offer.

Nonliteral interpretations of Genesis are not new. Some early Christians were already thinking this way. Genesis 1 contains obviously symbolic verses. The rest of the Bible contains figures of speech that are *not always* obvious. Yes, some verses suggest that the earth should be flat, and others do not. And it's not easy to tell which ones are poetry and which are not, unless you know from *another* source. Why should this be a problem? Both the Old Testament and the New Testament mention the "entire world" when we *now* know that the entire world was not literately involved. No Chinese came to Solomon, and Paul never preached in South America. So why should a "day" of Genesis be considered a literal twenty-four-hour day, especially when there's no sun to shine at all? Again, does it lessen the value of the Bible? By no means.

The world is more than six thousand years old, but it shouldn't be a threat to anyone's faith. It is not a theory; it is not a deduction from shaky hypotheses. It is an *observed* fact. Distances to stars farther away than six thousand light years have been geometrically measured. The light we receive from them demonstrates the laws of nature are the same over there as here on Earth. This light therefore had to travel for more than six thousand years in order to reach us.

Radio-dating techniques are often challenged (by young Earth creationists, for example) to prove that decay rates are stable with time. Astronomical observations show the laws of nuclear physics were the same tens of thousands years ago as now. Astrophysics has proved that nuclear physics had the same laws long ago. Decay rates, too, were the same. It has been *observed*.

The only coherent alternative to an old universe lies in a miraculous intervention from God. One option is that God could have created the universe with an *appearance* of age—starlight was created on its way to us, old trees came into existence with a number of rings already in place, polar caps were already a few kilometers thick when they appeared, and so on. As is sometimes argued, Jesus *instantly* created good wine from water, and *instantly* withered a fig tree, both of which take a certain amount of time according to natural laws. I tend to feel uncomfortable with this option because God seems to be deceiving us, and also because it is basically impossible to prove wrong (how does one prove that the universe was *not* created last Thursday, with the perfect appearance of age, including our memories?). But I will admit it is logically coherent.

There was a time when some believed that *even* if the earth was a sphere, no one would live at the antipodes. Based on his interpretation of Scriptures, Augustine wrote,

> As to the fable that there are Antipodes, that is to say, men on the opposite side of the Earth, where the Sun rises when it sets on us, men who walk with their feet opposite ours, there is no reason for believing it. Those who affirm it do not claim to possess any actual information; they merely conjecture that, since the Earth is suspended within the concavity of the heavens, and there is as much room on the one side of it as on the other, therefore the part which is beneath cannot be void of human inhabitants. They fail to notice that, even should it be believed or demonstrated that the world is round or spherical in form, it does not follow that the part of the Earth opposite to us is not completely covered with water, or that any conjectured dry land there should be inhabited by men. For Scripture, which confirms the truth of its historical statements by the accomplishment of its prophecies, teaches not falsehood; and it is too absurd to say that some men might have set sail from this side and, traversing the immense expanse of ocean,

have propagated there a race of human beings descended from that one first man.[1]

There was also a time when some faithful Christians, based on their understanding of the Bible, thought that the sun was revolving around the earth. Indeed, verses like Joshua 10:13—"So the sun stood still, and the moon stopped, till the nation avenged itself on its enemies, as it is written in the Book of Jashar. The sun stopped in the middle of the sky and delayed going down about a full day"—could be interpreted in this way. According to Irenaeus, bishop of Lugdunum[2] in the second century AD, "The sun also, who runs through his orbit in twelve months, and then returns to the same point in the circle..."[3] Centuries later, Martin Luther wrote [59], "A comet is a star that runs, not being fixed like a planet, but a bastard among planets. It is a haughty and proud star, engrossing the whole element, and carrying itself as if it were there alone." And Jean Calvin, in his *Commentaries on Psalms* [19], declared, "How could the earth hang suspended in the air were it not upheld by God's hand? By what means could it maintain itself unmoved, while the heavens above are in constant rapid motion, did not its Divine Maker fix and establish it?"

Based on their readings of the Bible, these great men claimed that the sun was running through its orbit, that comets are stars, that planets are fixed, that the earth remains unmoved while the heavens are in rapid motion, and that no one can live in New Zealand.

We now know that planets do turn around the sun and that comets are not stars. While the former should be known by everyone, it may be useful to remind regarding the latter that more than ten probes have been launched to comets. One of them, the Deep Impact probe, even *crashed* on comet Tempel 1 on July 4, 2005, sending us pictures shot up to ninety seconds before the crash.

So are Augustine, Luther, and Calvin discredited by these quotes? Absolutely *not*. Though erroneous, why should these claims cast any doubts on their theology, on their thinking, on their lives? I perfectly understand how someone could deduce from Joshua 10:13 that the sun turns around the earth. How would you know, from the Bible *alone*, that it is the other way around? And how would you know—again, from the Bible *alone*—that Isaiah 40:22 ("He sits enthroned above the *circle* of the earth") is literal,

1. *The City of God*, XVI, 9.
2. Present-day Lyon, France.
3. *Against Heresies*, Bk. 1, Ch. 27.

whereas Daniel 4:10–11 or Luke 4:5, which imply a flat Earth, are symbolic? Admittedly, there is no purely biblical way to know for sure that the earth is a sphere turning around the sun. But why should there be? Is it necessary for salvation? The Bible does not say that our sun is burning hydrogen, nor does it say that atoms are made of protons, neutrons, and electrons. The Bible is silent on scores of interesting issues. This silence is not a problem at all for the Bible. But it could be a problem for some rigid interpreters.

Suppose we take one of these believers of the past into space. He suddenly sees with his naked eye that our planet is a sphere and that people live all around it. Flying even farther into space, he sees that the earth, together with all the planets, do revolve around the sun.

How would he react? Sadly, he might go through a severe crisis of faith, possibly causing him to lose it. After all, he would think, doesn't the Bible teach that the earth is flat and fixed, while the sun revolves around it? I hope our believer would be flexible enough to admit that the problem was never the Bible, but his *interpretation* of the Bible. The Bible can be trusted. And the earth is neither flat nor fixed. How sad would it be to lose faith for this reason, when the vision of the earth from afar could be such a tremendous source of spiritual inspiration!

Indeed, in their book *Finding Faith, Losing Faith: Stories of Conversion and Apostasy* [63], Scot McKnight and Hauna Ondrey list confrontation with modern science among the main reasons why people today lose their faith.

In Greek mythology we read the story of Theseus, who sailed from Athens to Crete to kill the Minotaur. Before he left, his father, Aegeus, had him promise that upon his return he would use *white* sails to signal success. *Black* sails would signal that he had died in the attempt. Theseus did kill the Minotaur, but he forgot to raise white sails on his return to Athens. Sighting black sails from afar, Aegeus mistakenly thought that his son had died. His grief was so deep that he jumped into the sea[4] and drowned. For nothing.

In a similar way, thousands of people this year will commit spiritual suicide for nothing. They will read a book, watch a documentary, or go to college, and be confronted by the evidence for an old universe. Then they will remember the young universe theology they were taught and relegate the Bible to the level of an interesting fairy tale, at best. Thousands this year will lose their faith because no one has told them that both "God causes his sun to rise" and "the sun rises because the earth rotates" are *true*.

---

4. The Aegean Sea, by the way.

CONCLUSION 89

So, there was a time when some pretended that the Bible teaches that the earth is flat and fixed. It is now very difficult, though not impossible, to find people making the same argument—not because Christians are cowards intimidated by NASA, but simply because the verses that seem to support the notion of a flat earth can be considered metaphorical, figures of speech, without doing any harm to the message of the Bible. Why would anyone, against the evidence, hold on to a metaphor? While I wish the current proponents of a young universe would likewise surrender their position, and free themselves from their interpretation of the Bible, they need to realize that their attitude has serious consequences.

An allegoric reading of Genesis 1 does not harm the message of the Bible, because the text itself does not demand a literal reading (see chapter 2). But claiming that the world is six thousand years old, according to the Bible, does harm the Bible's purpose. How? Before he left, Jesus gave a mission to his followers: "go and make disciples of all nations" (Matt 28:19). The Bible does teach that the world is lost, that people need to be saved, that "no one comes to the Father except through me [Jesus]" (John 14:6), and that Christians, like Paul, are "to win as many as possible" (1 Cor 9:19).

Perhaps many people don't come to Jesus because they don't want to change their lives and don't want to "come into the light for fear that their deeds will be exposed" (John 3:20). But some don't come to Jesus or accept the teachings of the Bible because they have witnessed poor examples set by alleged Christians *before* they read the Bible, and they associate the latter with the former. Some have heard about sexual abuse by clergy members, or have seen how the Catholic Church walked hand in hand with ruthless dictators (like in Chile, Argentina, or Spain), and they have concluded that the Bible cannot be a good book if it countenances such actions. Still others, only slightly aware of the discoveries regarding the age of the universe, could have been truly fascinated by the Bible, if only they hadn't *first* heard someone pretend that it teaches a six-thousand-year-old universe. They have walked away thinking that they should turn to something more serious. Granted, the responsibility is still theirs to open the book one day, but Christians should draw people to the Bible instead of bringing "the way of truth into disrepute" (2 Pet 2:2). To quote St. Peter again, "If you suffer, it should not be as a murderer or thief or any other kind of criminal, or even as a meddler" (1 Pet 4:15).

Christians can be insulted for "good" reasons, but also for bad ones. Jesus himself was eventually put to death, so Christians should not expect to receive nothing but praise from society. But the point Peter makes here is that Christians should not be criticized for *bad* reasons. The way of truth should not be brought into disrepute for bad reasons. Making the claim that the Bible teaches a flat earth had no spiritual consequences six hundred years ago. But making the claim that the Bible teaches a six-thousand-year-old universe *today* does have negative spiritual consequences. Indeed, because of this false teaching, some will wander away, while others will be kept away. My hope is that this little book will help both groups.

Appendix A: A Little Atomic Physics

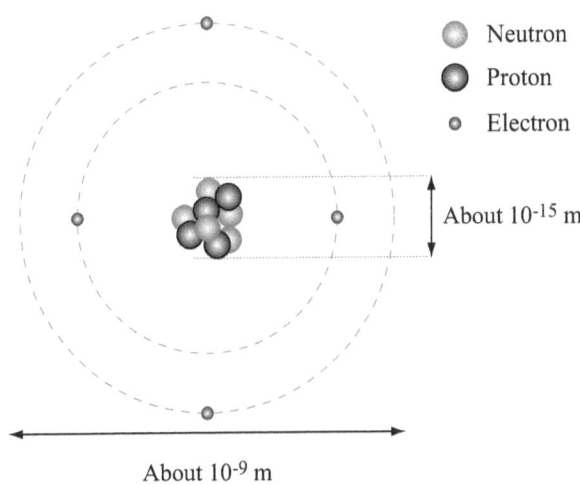

Figure 22: Naive picture of an atom. The size of the nucleus is about 10^{-15} m, that is, one preceded by fourteen zeros. The overall size of the atom is approximately 10^{-9} m, a million times larger than the nucleus. Needless to say, the picture is not to scale. With five neutrons, four protons, and four electrons, we have here a beryllium-9 atom.

Atoms are the building blocks of matter. Surprisingly, the idea that matter is made of elementary "bricks" is thousands of years old, as the Greek philosopher Democritus posed this hypothesis around 400 BC. Although the word *atom* comes from the Greek *atomos*, meaning "indivisible," we now know that atoms can be divided into protons, neutrons, and electrons. Electrons have been found to be indivisible, while protons and neutrons are not. The size of an atom is typically one billionth of a meter—you could literally line up one billion atoms in a single meter—which explains why matter around us seems continuous instead of discrete.

At the center of an atom lies its nucleus, which is composed of protons and neutrons. Around the nucleus, electrons orbit in equal number to the protons in the nucleus. An atom could be *very schematically* pictured as in figure 22 above. It is impossible to render the scale correctly, as the nucleus is about one million times *smaller* than the atom as a whole. If the nucleus was the size of a golf ball, the atom would be forty-two kilometers long (the length of a marathon). The way the electrons are arranged around the nucleus is described by a mathematical function $\Psi(r)$, solution of the Schrödinger equation:

$$E\psi(r) = -\frac{\hbar^2}{2m} \nabla^2 \psi(r) + V(r)\psi(r) \qquad (4)$$

Consider the simplest atom of all: the hydrogen atom. The hydrogen atom is formed by just one electron orbiting around one proton. When applied to this case, the equation above predicts that the electron has various ways of orbiting around the proton, which are called "levels." For each "level," it will have a different energy. When it goes from one level to another, the atom emits or absorbs one particle of light. If the initial level is higher in energy, the atom emits light. If lower, it absorbs light. The levels can be numbered, and thus labeled, using an integer n ranging from one to infinity. Though not very accurate, it is common to represent these levels as circles around the nucleus. Suppose the electron goes from the initial level n_i to the final level n_f, then the energy difference between the two levels is as follows:

$$\Delta E = 2\pi^2 \frac{m_e q_e^4}{h^2} \left(\frac{1}{n_i^2} - \frac{1}{n_f^2} \right) \qquad (5)$$

The name and the value of the various parameters are given in table 5 below. Note that although not immediately apparent (to say the least), formula (5) is a mathematical consequence of formula (4).

Parameters	Name	Value
m_e	Electron mass	$9.1 \cdot 10^{-31}$ kg
q_e	Electron charge	$1.6 \cdot 10^{-19}$ C
h	Planck's constant	$6.6 \cdot 10^{-34}$ J·s
c	Velocity of light	$3 \cdot 10^8$ m/s

Table 5: Name and value of the various parameters involved in formulas (5) and (6).

n_i	λ (nm)
3	656
4	486
5	434
6	410

Table 6: Wavelengths λ in nanometers (nm) obtained from formulas (5) and (6) with $n_f = 2$.

Now, we have seen that when the electron in the hydrogen atom loses energy, it gives it to a particle of light that escapes the atom. Labeling λ the wavelength of this particle, we know how to relate it to the energy lost. The formula for this is as follows:

$$\lambda = \frac{hc}{\Delta E} \qquad (6)$$

Here c is the speed of light given in table 5. We thus know which kind of light hydrogen atoms should emit. Combining formulas (5) and (6), we can deduce that hydrogen is going to emit a number of wavelengths that we can detect. Considering $n_f = 1$ and $n_i = 1, 2, 3 \ldots$ in formula (5) yields in turn through formula (6) a series of very definite wavelengths. Table 6 shows the wavelengths obtained when considering $n_f = 2$ and a few values of n_i. This series of wavelengths is called the Balmer series, after Johann Jakob Balmer, who was able to observe it in 1885 because these wavelengths are *visible*.

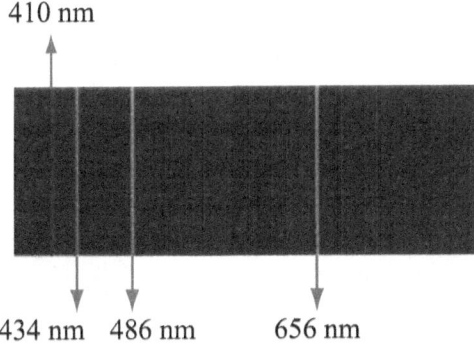

Figure 23: Visible hydrogen lines.

Nowadays, a typical practical exercise in a physics curriculum precisely consists in observing these wavelengths. Here is how the experience goes: students switch on a hydrogen lamp. This lamp is simply a tube, filled with hydrogen, in which a current circulates as in a neon lamp. The current heats the hydrogen so that the electrons around the protons start jumping up and down from one level to another, emitting light the wavelength of which fits formula (5). Students also have a spectrometer, an apparatus that can detect particles of light and measure their wavelengths. The spectrometer can detect visible and nonvisible light, but when focusing on the visible part of the emission, students observe something similar to what is depicted in figure 23.

The four lines predicted in table 6 are found exactly where expected. The lamp appears red because the line at 656 nm is the most intense one in the visible range. The intensity of this line, as well as every other line, can be predicted from formula (4). In the mid-nineteenth century, scientists were at first puzzled to observe such "lines of emission." They expected a continuum, where every possible wavelength would have been represented. This unexpected observation was one of the triggers of quantum mechanics, for which formula (4) stands as an icon.

The complete set of wavelengths that hydrogen can emit is called its spectrum. Each atom has its own spectrum. Each molecule has its own spectrum. Emission lines from different atoms or molecules never overlap exactly. That means that if we study the light coming from the sun with a spectrometer, we see all those lines, and because lines from different atoms don't overlap, we can tell which atoms are in the sun by looking at the emission lines. This is how, as early as 1862, Anders Jonas Ångström found that there was hydrogen in the sun. Although quantum mechanics had not been devised yet, he observed that the lines from a hydrogen discharge were present in the sunlight and deduced that the same hydrogen was sending them from the sun.

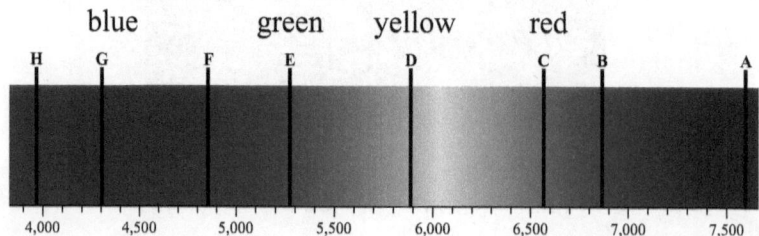

Figure 24: A portion of the sun's spectrum. Color goes from blue on the left side, to red around letter C, through green (E) and yellow (D). The absorption of light around letters C and F (among others), at 656.3 and 486.1 nm respectively, indicates that there is hydrogen in the sun. Graduation is in angstroms (Å), with 6,563 Å = 656.3 nm.

Hydrogen can emit light of a given color when jumping from level two to level one, but it can also *absorb* the same kind of light when jumping from level one to level two. Depending on the circumstances, hydrogen will be detected either by the emission of some color, or by the *absorption* of the same color. Indeed, this is how hydrogen is detected in the sun. Figure 24 shows a portion of the sun's spectrum. Look at the black bar below the letter C. It is located at 6563 Å, that is, 656.3 nm. If you now go to table 6, you will see that this wavelength pertains to the transition from one level to another of the hydrogen atom. From this simple dark line, we can tell that there is hydrogen in the sun. This color is unique to hydrogen. And the other lines tell us about other atoms. Lines A and B, for example, betray the presence of oxygen up there, line D pertains to sodium, and so on.

In fact, not only does every atom have its unique spectrum, but every assembly of atoms, every molecule, also has its distinctive spectrum. For example, the water molecule (H_2O), made of two hydrogen atoms and one oxygen, can vibrate in special ways according to the equations we've seen so far. According to one of these vibration modes, it can emit light at a frequency of 22.2 GHz and a wavelength of 1.35 cm. It is interesting to compare this value of 1.35 cm with the wavelengths in table 6 and figure 24. Atomic spectral lines are found around a few hundred nanometers, that is, around a few hundred thousandths of a centimeter! There's no way one can take an atom for a molecule, and vice versa. Molecular wavelengths are literally hundreds of thousands times *larger* than atomic ones. And yes, we have instruments to detect each one of these, so that by looking at something and checking its spectral lines, we can tell a lot about its atomic and molecular composition.

By analyzing the light coming from astronomical objects, it is possible to tell which atoms or molecules are in it. More important, observing these lines allows knowing whether formula (4) is still valid in the object. If we look at a star four thousand light years away and find the very same hydrogen lines, the very same helium lines, and so on, it means quantum mechanics is valid there as it is on earth. But there's even more: these lines tell us that the laws of nature *were* the same when light was emitted as they are now. This is not the application of present knowledge to the past. Here, we literally *see* the past.

Appendix B: A Little Nuclear Physics

An atom is formed by a nucleus of protons and neutrons, around which electrons are orbiting (see figure 22 page 91). The type of atom is determined by the number of protons and neutrons in its nucleus—nothing else. The biggest one we have discovered contains some 110 protons and 170 neutrons. Well-known atoms are hydrogen, with one proton and one neutron in its nucleus; oxygen, with eight protons and neutrons; and carbon, with six protons and neutrons. It takes two atoms of hydrogen and one atom of oxygen to form a molecule we call water! By combining zero to 110 protons with one to 170 neutrons, we can form all the nuclei listed in figure 25. This table of nuclides lists every single atom that has ever been observed. Beyond the region covered by the squares, nuclei are so unstable that they have never been detected. It does not mean they can't exist. What is sure is that far from the borders, nuclei can't exist at all. Start from a still existing one out of the border and try to feed it with one more neutron or proton. It will just refuse to take it.

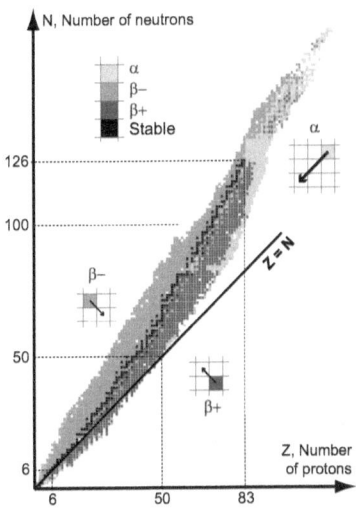

Figure 25: The table of nuclides. Each cell of the graph represents an atom. Only black cells are stable.

Every single cell of that grid has been studied individually. By googling "table of nuclides," you will find many sites where you can look at that grid, zoom in on any given cell, and find out everything about it.[1]

Atoms can be stable or unstable. Take a bunch of *stable* atoms and look at them: they will stay there forever. As you can see, the black squares, which represent stable atoms, all lie near the center of the domain. Take a bunch of *unstable* atoms and observe them: you'll see their number decreasing. Some spontaneously decay to form other atoms, which can be stable or unstable. If you wait long enough, from a few billionths of a second to millions of years, depending on the kind of atom you consider, there will be nothing left. Every single unstable atom that was there initially eventually decayed to something else. This phenomenon *is* radioactivity.

Note that I use the words *atoms* and *nucleus* interchangeably. This is because the atom is nothing other than its nucleus, plus some electrons orbiting around it. There is therefore a strict one-to-one correspondence between an atom and its nucleus.

Now, if an atom is unstable and becomes another, this other *must be* in the table. There is nowhere else to go. So radio decay is just a jump from one square to another. Here are the main ways in which an unstable atom can decay:

- α decay: The atom ejects two protons and two neutrons, that is, a helium nucleus.
- β- decay: A neutron in the nucleus changes into a proton. An electron and an antineutrino[2] are ejected.
- β+ decay: A proton in the nucleus changes into a neutron. A positron and a neutrino are ejected.
- Fission: The nucleus splits in two parts (not shown in figure 25).

If you look at figure 25, you'll see there is a logic in the way unstable atoms decay on the chessboard. Depending on where they start from, they

---

1. Try this site: www.nndc.bnl.gov/chart/ or www-nds.iaea.org/relnsd/vchart/.
2. The neutrino is the particle that was once thought to travel faster than light. The antineutrino and positron are the *anti*-particles of the neutrino and electron, respectively. *Anti*-matter is not science fiction. It has been routine in the lab for more than fifty years.

move so as to get closer to a stable cell. This is why all the β- unstable atoms are *above* the stable region, and all the β+ *below*. Because the heaviest stable atom has 126 neutrons and 83 protons, atoms heavier than that need a quick diet. In nuclear physics, that is the α decay, when you move down the chessboard two cells in each direction. As you can see, the landscape is not clear-cut for these heavy nuclei, as they can decay in almost every way. Fission is *the* radical diet: some heavy nuclei just split into two lighter elements located in the lower part of the table. Nuclear power plants rely on this process, as fission events release energy.

The straight line on the table of nuclides indicates the cells with an even number of protons and neutrons. Interestingly, stable nuclei have more neutrons than protons. This is because protons tend to repel each other, while neutrons stick to other neutrons and to protons. Putting more neutrons than protons provides therefore a glue helping the protons bear with each other. But enough is enough. Beyond 83 protons, no stable nuclei have been found so far. Past this point, there's no way neutrons can help.

All decay events but fission force the atom to switch to something only slightly different. Like the pawn or the knight on the chessboard, non-fission decays have the atom move only one or two cells. But with fission, we have the equivalent of a queen. The atom jumps about halfway down the grid and, indeed, the fragments don't have anything to do with the original.

Count the cells in figure 25. You'll find about 3,200 nuclides, only 260 of which are stable. Every single one of these nearly three thousand unstable atoms decays according to the very same law.[3] Assume you start with a number of parent atoms. Wait a time τ, which depends on the atom and on the decay mode, and half the parents turn into daughters. Wait another time τ, and again half of the remaining parents turn into daughters. And so on, until all the parents are gone. This is at the heart of radio-dating techniques.

---

3. Reference [34] gives a nice overview of the "nuclear landscape."

Why Do α Decay Rates Not Vary?

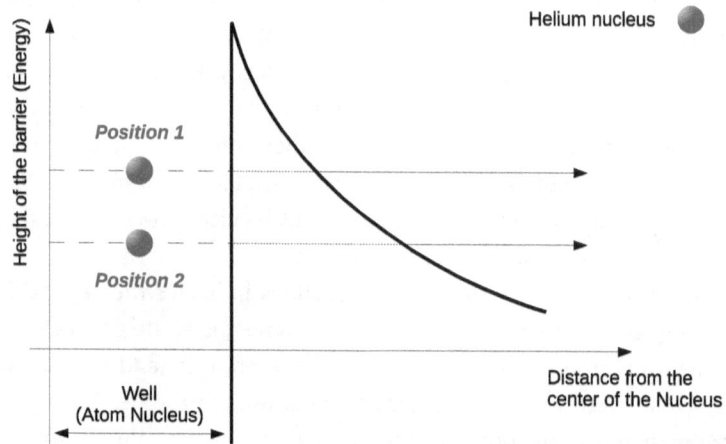

Figure 26: How α decay works. The helium inside the atom nucleus leaves it, passing through the barrier. Probability of leaving successfully is higher from position 1 than from position 2. If the picture were to scale, the orbiting electrons would be found some twenty kilometers to the right.

Chapter 5 reports how decay rates have been found to vary when modifying the electronic cloud around the nucleus. Interestingly, among all the possible decay modes, α decay rates don't vary. Why are the elements for which the decay rate is expected to potentially vary all β emitters? Because this kind of decay involves the atom's electronic cloud, whereas α decay does not. α decay consists in ejecting a helium nucleus (two protons and two neutrons) at once. Before it was ejected, this helium was inside the unstable nucleus. It was trapped, in exactly the same way you could be trapped in the well depicted in figure 26. If you could climb high enough to get out of the well, your descent would be easy because once you're out, your two protons repel with the protons from the nucleus.[4] This is why there is a slope after the well.

But here, you can't climb because that requires energy, and there's nowhere you can get it from. Nevertheless, there is a way out. Quantum mechanics, through equation (4) page 92 (plus some heavy math), allows you to pass *through* the barrier and escape the well just with the energy you

---

4. The force repelling particles of the same charge is called the Coulomb force. Inside the nucleus, protons still Coulomb-repel. But because there are so close to each other there, another kind of force, the so-called "strong nuclear force" can operate. An attraction results, completely overcoming Coulomb repulsion.

have. This is called the quantum tunnel effect (previously mentioned on page 28).

From the laws we know, the odds of leaving the nucleus within a certain amount of time can be calculated. The Russian physicist George Gamow was the first one who thought about all this in 1928. He derived a formula that fit very well with the empirical Geiger-Nuttall law that was known at the time for α decay. It just depends on the shape of the well and of the slope outside, and of the initial position of the helium inside the nucleus. Position 1 is easier to exit from than position 2, simply because the barrier is thinner.

Where are the orbiting electrons in that picture? Approximately one million times beyond the border of the page. So if the picture were to scale, you would find them about twenty kilometers to the right. This is why chemistry, and everything you could do to these electrons, has a *very* hard time changing α decay rates. It all happens in another world.

Bibliography

[1] Y. Amelin, et al. Lead isotopic ages of chondrules and calcium-aluminum-rich inclusions. *Science*, 297:1678, 2002.
[2] T. Aoyama, et al. Tenth-Order QED Contribution to the Electron g-2 and an Improved Value of the Fine Structure Constant. *Physical Review Letters*, 109:111807, 2012.
[3] J. R. Arnold and W. F. Libby. Age determinations by radiocarbon content: Checks with samples of known age. *Science*, 110:678, 1949.
[4] B. Aschenbach. Discovery of a young nearby supernova remnant. *Nature*, 396:141, 1998.
[5] D. R. Atanasov, et al. Half-life measurements of stored fully ionized and hydrogen-like I-122 ions. *European Physical Journal A*, 48:22, 2012.
[6] R. Barjavel. *Ashes, Ashes.* New York: Curtis, 1971.
[7] E. Bellotti, et al. Search for time dependence of the 137Cs decay constant. *Physics Letters B*, 710:114, 2012.
[8] P. Benoist, et al. The decay probability of Be-7 as a function of the ionization of the atom. *Physical Review*, 76:1000, 1949.
[9] D. W. Bercot, editor. *A Dictionary of Early Christians Beliefs.* Peabody, MA: Hendrickson, 1998.
[10] A. Biswas and K. Mani. Relativistic perihelion precession of orbits of Venus and the Earth. *Central European Journal of Physics*, 6:754, 2008.
[11] X.-J. Bi, et al. Constraints and tests of the opera superluminal neutrinos. *Physical Review Letters*, 107:241802, 2011.
[12] F. Bosch, et al. Observation of bound-state β- decay of fully ionized Re187: Re187/Os187 cosmochronometry. *Physical Review Letters*, 77:5190, 1996.
[13] C. F. Bradshaw, et al. High-resolution parallax measurements of Scorpius X-1. *The Astrophysical Journal Letters*, 512:L121, 1999.
[14] W. F. Brisken, et al. Very long baseline array measurement of nine pulsar parallaxes. *The Astrophysical Journal*, 571:906, 2002.
[15] E. Brook. Palaeoclimate: Windows on the greenhouse. *Nature*, 453:291, 2008.
[16] G. Brumfiel. The race to break the standard model. *Nature*, 455:156, 2008.
[17] G. Brumfiel. Hint of Higgs, but little more. *Nature*, 475:434, 2011.
[18] A. Brunthaler, et al. Trigonometric Parallaxes of Massive Star-Forming Regions. V. G23.01-0.41 and G23.44-0.18. *The Astrophysical Journal*, 693:424, 2009.
[19] J. Calvin. *Commentary on Psalm.* Vol. 4. Psalm 93.1.
[20] S. Chatterjee, et al. Precision astrometry with the very long baseline array: Parallaxes and proper motions for 14 pulsars. *The Astrophysical Journal*, 698:250, 2009.
[21] S. Chatterjee, et al. Getting its kicks: A vlba parallax for the hyperfast pulsar b1508+55. *The Astrophysical Journal Letters*, 630:L61, 2005.

[22] A. Cho. Higgs boson makes its debut after decades-long search. *Science*, 337:141, 2012.

[23] G. M. Clemence. The relativity effect in planetary motions. *Review of Modern Physics*, 19:361, 1947.

[24] C. Codella, et al. Looking for high-mass young stellar objects: H2O and OH masers in ammonia cores. *Astronomy & Astrophysics*, 510:A86, 2010.

[25] J. A. Cooper, et al. Effect of the chemical state on the lifetime of the 24-second isomer of nb-90. *Physical Review Letters*, 15:680, 1965.

[26] R. Cowsik, et al. Superluminal neutrinos at opera confront pion decay kinematics. *Physical Review Letters*, 107:251801, 2011.

[27] T. M. Dame, et al. The milky way in molecular clouds: A new complete CO survey. *The Astrophysical Journal*, 547:792, 2001.

[28] W. Dansgaard, et al. Evidence for general instability of past climate from a 250-kyr ice-core record. *Nature*, 364:218, 1993.

[29] A. T. Deller, et al. Precision southern hemisphere pulsar vlbi astrometry: Techniques and results for psr j1559-4438. *The Astrophysical Journal*, 690:198, 2009.

[30] R. Diehl, et al. Al-26 in the inner Galaxy. Large-scale spectral characteristics derived with SPI/INTEGRAL. *Astronomy & Astrophysics*, 449:1025, 2006.

[31] A. Einstein and L. Infeld. *The Evolution of Physics: The Growth of Ideas from Early Concepts to Relativity and Quanta*. New York: Simon & Schuster, 1961.

[32] G. T. Emery. Perturbation of nuclear decay rates. *Annual Review of Nuclear Science*, 22:165, 1972.

[33] A. G. Engelkemeir, et al. The half-life of radiocarbon (c14). *Physical Review*, 75:1825, 1949.

[34] J. Erler, et al. The limits of the nuclear landscape. *Nature*, 486:509, 2012.

[35] G. Feinberg. Possibility of faster-than-light particles. *Physical Review*, 159:1089, 1967.

[36] R. Feynman. *The Pleasure of Finding Things Out: The Best Short Works of Richard P. Feynman*. New York: Basic Books, 2005.

[37] A. V. Filippenko. Optical spectra of supernovae. *Annual Review of Astronomy and Astrophysics*, 35:309, 1997.

[38] E. Fischbach, et al. Time-dependent nuclear decay parameters: New evidence for new forces? *Space Science Reviews*, 145:285, 2009.

[39] E. Fischbach, et al. Evidence for time-varying nuclear decay rates: Experimental results and their implications for new physics. *arXiv:1106.1470*, June 2011.

[40] J. R. Forster and J. L. Caswell. Radio continuum emission at OH and H2O maser sites. *The Astrophysical Journal*, 530:371, 2000.

[41] D. Galenson. *White Servitude in Colonial America: An Economic Analysis*. Cambridge: Cambridge University Press, 1984.

[42] D. R. Gies and C. T. Bolton. The optical spectrum of HDE 226868 = Cygnus X-1. II Spectrophotometry and mass estimates. *The Astrophysical Journal*, 304:371, 1986.

[43] G. F. Giudice, et al. Interpreting opera results on superluminal neutrino. *Nuclear Physics B*, 861:1, 2012.

[44] H. Griem. *Principles of Plasma Spectroscopy*. Cambridge Monographs on Plasma Physics. Cambridge: Cambridge University Press, 2005.

[45] H. Gupta, editor. *Encyclopedia of Solid Earth Geophysics*. Encyclopedia of Earth Sciences Series. Dordrecht: Springer, 2011.

[46] J. C. Hafele and R. E. Keating. Around-the-world atomic clocks: Predicted relativistic time gains. *Science*, 177:166, 1972.

[47] D. Hanneke, et al. New measurement of the electron magnetic moment and the fine structure constant. *Physical Review Letters*, 100:120801, 2008.

[48] R. W. Hanuschik and T. Schmidt-Kaler. Absorption line velocities and the distance of supernova 1987a. *Astronomy & Astrophysics*, 249:36, 1991.

[49] J. C. Hardy, et al. Do radioactive half-lives vary with the Earth-to-Sun distance? *ArXiv:1108.5326*, 2011.

[50] V. Hugo and H. Juin. *Choses vues: souvenirs, journaux, cahiers, 1830–1885*. Quarto (Paris): Gallimard, 2002.

[51] A. F. Iyudin, et al. Comptel observations of Ti-44 gamma-ray line emission from Cas-A. *Astronomy & Astrophysics*, 284:L1, 1994.

[52] J. Jouzel, et al. Orbital and millennial Antarctic climate variability over the past 800,000 years. *Science*, 317:793, 2007.

[53] K. Kretschmer, et al. Al-26 spectroscopy with SPI: The challenge to detect Galactic rotation. *Advances in Space Research*, 38:1439, 2006.

[54] M. J. Kuchner, et al. Evidence for Ni-56 yields Co-56 yields Fe-56 decay in type IA supernovae. *The Astrophysical Journal Letters*, 426:L89, 1994.

[55] L. Landau and E. Lifshitz. *The Classical Theory of Fields*. Translated by Morton Hamermesh. Course of Theoretical Physics 2. Oxford: Butterworth-Heinemann, 1975.

[56] Leloudas, et al. The normal type IA sn 2003hv out to very late phases. *Astronomy & Astrophysics*, 505:265, 2009.

[57] Y. A. Litvinov and F. Bosch. Beta decay of highly charged ions. *Reports on Progress in Physics*, 74:016301, 2011.

[58] L. Loulergue, et al. Orbital and millennial-scale features of atmospheric CH_4 over the past 800,000 years. *Nature*, 453:383, 2008.

[59] M. Luther. *Table Talks*. Chapter 124.

[60] D. Lüthi, et al. High-resolution carbon dioxide concentration record 650,000-800,000 years before present. *Nature*, 453:379, 2008.

[61] S. M. Matz, et al. Gamma-ray line emission from SN1987a. *Nature*, 331:416, 1988.

[62] K. McGuffie and A. Henderson-Sellers. *A Climate Modelling Primer*. 3rd ed. Research and Developments in Climate. Hoboken, NJ: Wiley, 2005.

[63] S. McKnight and H. Ondrey. *Finding Faith, Losing Faith: Stories of Conversion and Apostasy*. Waco: Baylor University Press, 2008.

[64] Z. Merali. Collider gets yet more exotic 'to-do' list. *Nature*, 466:426, 2010.

[65] M. Messineo, et al. 86 Ghz SiO maser survey of late-type stars in the inner galaxy. i. observational data. *Astronomy & Astrophysics*, 393:115, 2002.

[66] G. A. Moellenbrock, et al. A precise distance to iras 00420+5530 via $H2O$ maser parallax with the vlba. *The Astrophysical Journal*, 694:192, 2009.

[67] L. Moscadelli, et al. Trigonometric parallaxes of massive star-forming regions. ii. cep a and ngc 7538. *The Astrophysical Journal*, 693:406, 2009.

[68] Z. Ninkov, et al. The primary orbit and the absorption lines of hde 226868 (Cygnus X-1). *The Astrophysical Journal*, 321:425, 1987.

[69] E. B. Norman, et al. Evidence against correlations between nuclear decay rates and earth sun distance. *Astroparticle Physics*, 31:135, 2009.

[70] N. Panagia, et al. Properties of the SN1987A circumstellar ring and the distance to the large magellanic cloud. *The Astrophysical Journal Letters*, 380:L23, 1991.

[71] M. R. Pestalozzi, et al. A general catalogue of 6.7-ghz methanol masers. i. data. *Astronomy & Astrophysics*, 432:737, 2005.

[72] O. Pétré-Grenouilleau. *Les Traites négrières: Essai d'histoire globale.* Collection Folio Histoire. Paris: Gallimard, 2006.

[73] M. C. Pinheiro, et al. The young stellar cluster [DBS2003] 157 associated with the H II region GAL 331.31-00.34. *Monthly Notices of the Royal Astronomical Society*, 423:2425, 2012.

[74] A. Ray, et al. Decay rate of beryllium-7 in different environments. *Science*, 287:1203, 2000.

[75] J. E. Reed, et al. The three-dimensional structure of the Cassiopeia A supernova remnant. i. the spherical shell. *The Atrophysical Journal*, 440:706, 1995.

[76] M. J. Reid, et al. The trigonometric parallax of Cygnus X-1. *The Astrophysical Journal*, 742:83, 2011.

[77] M. J. Reid, et al. Trigonometric parallaxes of massive star-forming regions. i. s 252 & g232.6+1.0. *The Astrophysical Journal*, 693:397, 2009.

[78] P. Reimer, et al. IntCal04 terrestrial radiocarbon age calibration, 0-26 cal kyr BP. *Radiocarbon*, 46:1029, 2004.

[79] J. Z. Ren, et al. Outflow activities in the young high-mass stellar object g23.44-0.18. *Monthly Notices of the Royal Astronomical Society*, 415:L49, 2011.

[80] A. Sanna, et al. Trigonometric parallaxes of massive star-forming regions. ix. the outer arm in the first quadrant. *The Astrophysical Journal*, 745:82, 2012.

[81] A. Sanna, et al. Trigonometric parallaxes of massive star-forming regions. vii. g9.62+0.20 and the expanding 3 kpc arm. *The Astrophysical Journal*, 706:464, 2009.

[82] M. Sato, et al. Trigonometric parallax of w51 main/south. *The Astrophysical Journal*, 720:1055, 2010.

[83] V. Schönfelder, et al. Ti-44 gamma-ray line emission from cas a and rxjo852-4622/grojo852-4642. In *The Fifth Compton Symposium*, edited by M. L. McConnell and J. M. Ryan, 54. American Institute of Physics Conference Proceedings 510. Melville, NY: American Institute of Physics, 2000.

[84] N. S. Schulz, et al. The first high-resolution x-ray spectrum of cygnus x-1: Soft x-ray ionization and absorption. *The Astrophysical Journal*, 565:1141, 2002.

[85] E. Segrè and C. E. Wiegand. Experiments on the effect of atomic electrons on the decay constant of be-7. *Physical Review*, 75:39, 1949.

[86] J. Speer. *Fundamentals of Tree-Ring Research.* Tucson: University of Arizona Press, 2010.

[87] D. Steeghs and J. Casares. The mass donor of Scorpius X-1 revealed. *The Astrophysical Journal*, 568:273, 2002.

[88] M. Stuiver, et al. INTCAL98 radiocarbon age calibration, 24,000-0 cal BP. *Radiocarbon*, 40:1041, 1998.

[89] M. Stuiver, et al. High-precision radiocarbon age calibration for terrestrial and marine samples. *Radiocarbon*, 40:1127, 1998.

[90] K. Takahashi, et al. Bound-state beta decay of highly ionized atoms. *Physical Review C*, 36:1522, 1987.

[91] E. Taylor and J. Wheeler. *Exploring Black Holes: Introduction to General Relativity.* San Francisco: Addison Wesley Longman, 2000.

[92] R. Thom. *René Thom (1923-2002).* Paris: Société mathématique de France, 2005.

[93] J. N. Tournier. *Le vivant décodé : Quelle nouvelle définition donner à la vie?* Les Ulis, France: EDP Sciences, 2005.

[94] J. Uzan and B. Leclercq. *The Natural Laws of the Universe: Understanding Fundamental Constants.* Springer Praxis Books in Popular Astronomy. Berlin: Springer, 2008.

[95] G. F. Varani, et al. Direct observation of radioactive cobalt decay in supernova 1987a. *Monthly Notices of the Royal Astronomical Society*, 245:570, 1990.

[96] A. J. Walsh, et al. Studies of ultracompact HII regions - II. high-resolution radio continuum and methanol maser survey. *Monthly Notices of the Royal Astronomical Society*, 301:640, 1998.

[97] W. Wang, et al. Spi observations of the diffuse 60Fe emission in the galaxy. *Astronomy & Astrophysics*, 469:1005, 2007.

[98] J. K. Webb, et al. Indications of a spatial variation of the fine structure constant. *Physical Review Letters*, 107:191101, 2011.

[99] A. Weiß, et al. On the variations of fundamental constants and active galactic nucleus feedback in the quasi-stellar object host galaxy rxj0911.4+0551 at $z = 2.79$. *The Astrophysical Journal*, 753:102, 2012.

[100] Y. Xu, et al. Trigonometric parallaxes of massive star-forming regions. viii. g12.89+0.49, g15.03-0.68 (m17) and g27.36-0.16. *The Astrophysical Journal*, 733:25, 2011.

[101] Y. Xu, et al. Trigonometric parallaxes of massive star-forming regions: Iii. g59.7+0.1 and w 51 irs2. *The Astrophysical Journal*, 693:413, 2009.

[102] D. Young and R. Stearley. *The Bible, Rocks and Time: Geological Evidence for the Age of the Earth*. Downers Grove, IL: InterVarsity, 2008.

[103] B. Zhang, et al. Trigonometric parallaxes of massive star-forming regions. iv. g35.20-0.74 and g35.20-1.74. *The Astrophysical Journal*, 693:419, 2009.

[104] "Editorial: No Shame." *Nature*, 484:287, 2012.

www.ingramcontent.com/pod-product-compliance
Lightning Source LLC
Chambersburg PA
CBHW030903170426
43193CB00009BA/729